QUANTUM MECHANICS, A HALF CENTURY LATER

W0106782

EPISTEME

A SERIES IN THE FOUNDATIONAL,

METHODOLOGICAL, PHILOSOPHICAL, PSYCHOLOGICAL,

SOCIOLOGICAL AND POLITICAL ASPECTS

OF THE SCIENCES, PURE AND APPLIED

Editor: MARIO BUNGE
Foundations and Philosophy of Science Unit, McGill University

VOLUME 5

QUANTUM MECHANICS,
A HALF CENTURY LATER

PAPERS OF A COLLOQUIUM
ON FIFTY YEARS OF QUANTUM MECHANICS,
HELD AT THE UNIVERSITY LOUIS PASTEUR,
STRASBOURG, MAY 2–4, 1974

Edited by

JOSÉ LEITE LOPES

Université Louis Pasteur, Strasbourg

and

MICHEL PATY

Centre de Recherches Nucleaires et Université Louis Pasteur, Strasbourg

D. REIDEL PUBLISHING COMPANY

DORDRECHT-HOLLAND/BOSTON-U.S.A.

Library of Congress Cataloging in Publication Data

Colloquium on Fifty Years of Quantum Mechanics,
    University Louis Pasteur, 1974.
    Quantum mechanics.

    (Episteme ; v. 5)
    Bibliography: p.
    Includes index.
    1. Quantum theory—Congresses. I. Lopes, José
Leite. II. Paty, Michel, 1938–   III. Title.
QC173.96.C64 1974   530.1'2   77–8178
ISBN-13:978-94-010-1198-3   e-ISBN-13:978-94-010-1196-9
DOI: 10.1007/978-94-010-1196-9

Published by D. Reidel Publishing Company,
P.O. Box 17, Dordrecht, Holland

Sold and distributed in the U.S.A., Canada, and Mexico
by D. Reidel Publishing Company, Inc.
Lincoln Building, 160 Old Derby Street, Hingham,
Mass. 02043 U.S.A.

TABLE OF CONTENTS

FOREWORD

The articles collected in this volume were written for a Colloquium on Fifty Years of Quantum Mechanics which was held at the University Louis Pasteur of Strasbourg on May 2–4, 1974, in commemoration of the original work by De Broglie in 1924.

It is our hope that this volume will convey to the reader the idea that quantum mechanics, besides being a fundamental tool for scientific workers today, is also a source of a number of questions and thoughts about the interpretation of the foundation of quantum mechanics itself. This gives rise to problems of a philosophical and logical character and has repercussions on other domains such as the theory of gravitation.

Besides the papers presented at the Colloquium, an article has been included by D. Bohm and B. Hiley. This compensates, perhaps, for the article of S. Kochen, whose manuscript unfortunately did not reach us in time for inclusion in this volume. A few months after this Colloquium we learned of the death of Professor Jauch, who had taken a lively and crucial part in its discussions. We have been extremely saddened by the news of his death, and would like to express our long standing indebtedness to him as a physicist.

We are grateful to Professor B. d'Espagnat who kindly helped us in organizing the Colloquium meetings and to Professor G. Ourisson who, as President of the Louis Pasteur University, gave us encouragement and support to our enterprise. We would further like to express our thanks to all those who have contributed to the work involved in the Colloquium and the publication of this book, and especially to Dr J. Simmons who agreed to check the English version of several contributions.

J. LEITE LOPES
M. PATY

PREFACE

A few years ago, H. Barreau, G. Monsonego, M. Paty and myself organized the Seminar on the Foundations of Science at the University Louis Pasteur of Strasbourg.

Modern society recognizes the importance of science in the world through its technological applications and the multitude of appliances, gadgets and mechanisms which fill and often invade our daily life.

Scientists, dominated by the spirit of competition in their respective fields of specialization, often pass too quickly from one interpretation scheme to another and thus may stifle their research work by being over-anxious to publish more and more.

It would therefore be beneficial to organize in the universities, parallel to the different specialized seminars on recent experimental work and theoretical models, seminars on the foundation of science, its historical evolution, and the description and genesis of its ideas.

That is what we have tried to set up at Strasbourg over the past few years; it is with this motivation that the Conference 'A Half Century of Quantum Mechanics' was organized, and these papers collected and published.

It is already fifty years ago that the intuition of De Broglie laid the foundation stone of the theoretical structure that houses the fundamental laws of atomic, nuclear and sub-nuclear phenomena.

As everybody knows, some fundamental notions of quantum mechanics put forward by De Broglie himself and by Einstein and Niels Bohr, amongst others, are still open to further discussion. It is true that perhaps a great majority of physicists remain content in calculating the effective cross-sections of collisions and reactions and the energy levels of physical systems etc... with the help of quantum mechanics. However, others try in addition to examine thoroughly its physical and philosophical significance, and are sometimes led to modify some basic ideas of the theory.

Such discussions may indeed contribute to the progress of scientific thought. We hope that this collection of papers which has been requested on the occasion of the 50th anniversary of the thesis of Louis De Broglie, will contribute to that aim.

We regret the recent death of Leon Rosenfeld, the eminent physicist, who was one of the prominent members of the Copenhagen school, and who came several times to Strasbourg to participate in the seminars on the Foundations of Science.

Let us dedicate this work to the memory of L. Rosenfeld.

J. Leite Lopes

JOHN ARCHIBALD WHEELER*

INCLUDE THE OBSERVER IN
THE WAVE FUNCTION?

ABSTRACT. The classical dynamics of Einstein's closed universe (idealized for simplicity to be empty except as excited by gravitational waves) is analyzed in no way more economically than by the standard Hamilton-Jacobi method. In it the essential ideas are (1) a spacelike 3-geometry, $^{(3)}\mathscr{G}$; (2) an infinite-dimensional superspace $\mathscr{S}$, each point (or collection of equivalent points) of which stands for one and only one $^{(3)}\mathscr{G}$; (3) a Hamilton-Jacobi function S (dimensions of action) or wave-phase $S/\hbar$ (measured in radians) that propagates through superspace, $S = S(^{(3)}\mathscr{G})$, according to the standard Einstein-Hamilton-Jacobi equation of A. Peres; (4) a 'constructive interference' between this family of wave crests and other families of wave crests (obeying the same EHJ equation) that picks out (U. Gerlach) 'Yes' points (constructive interference) in superspace from 'No' points (destructive interference); (5) definition through these 'Yes' points of a 'leaf of history' slicing through superspace; and (6) identification of this leaf of history with a particular spacetime $^{(4)}\mathscr{G}$ that satisfies Einstein's field equation. Thus, every spacelike slice through this $^{(4)}\mathscr{G}$ is a 'Yes' $^{(3)}\mathscr{G}$, and conversely. When one goes from classical theory to quantum theory, one goes from $S(^{(3)}\mathscr{G})$ to $\psi(^{(3)}\mathscr{G})$, with the approximate correspondence:

$$\psi \sim \text{(slowly varying amplitude)} \exp{(iS/\hbar)}.$$

When one enriches the content of the theory from pure Einstein geometrodynamics to geometry plus particle-fields and other fields, one still has every degree of freedom of the universe included in the independent variables upon which ψ depends. Therefore the observer himself would seem to be included in ψ. However, so to 'include the observer in' makes insuperable difficulties of principle, which are described. One has to accept that physics never permits a complete prediction, but only a forecast of a correlation between events (E. P. Wigner). One is led to recognize that a wave function 'encompassing the whole universe' is an idealization, formalistically perhaps a convenient idealization, but an idealization so strained that it can be used only in part in any forecast of correlations that makes physical sense. For making sense it seems essential most of all to 'leave the observer out of the wave function'.

1. PRELIMINARY SURVEY OF THE ISSUE

Physics treats everything as a dynamic system, even the universe. Moreover, it describes the state of every dynamic system, not deterministically, but in probability terms, by a state function. Thus the consistent carrying out of the scheme of physics as one understands it today leaves no alternative but to assign a probability

J. Leite Lopes and M. Paty (eds.), Quantum Mechanics, a Half Century Later, 1-18. All Rights Reserved.
Copyright © 1977 by D. Reidel Publishing Company, Dordrecht-Holland

amplitude to the state of the universe. Yet, Wigner reasons, there is no such thing as the state function of the universe. A state function is only useful insofar as it describes correlations between observations – and there is no place for any 'observer' to stand 'outside' the universe to observe it. There has to be a wave function for the universe but there can't be a wave function for the universe: that is the dilemma.

How does the 'dilemma of a wave function for the universe' arise? Why is it important? What changed view does it suggest of the nature of physics? Geometrodynamics? These issues about the reasonableness – and unreasonableness – of a 'quantum mechanics of the universe' are of concern today. There is no better occasion to survey them than this 50th anniversary of Louis de Broglie's discovery of the connection between quantum states and waves.

2. THE ISSUE MORE SHARPLY FOCUSSED: DOES SOME RELEVANT FEATURE OF CONSCIOUSNESS FALL OUTSIDE THE DOMAIN OF PHYSICS?

The methods of quantum mechanics allow us to treat the machinery of observation. Do they also allow us to analyze the observer?

We have no trouble of quantum-mechanics principle to apply the idea of 'wave function' when a particle passes through a cloud chamber. Nor when that particle ionizes a nitrogen molecule. Nor at later stages down the road: ion nucleating droplet; droplet scattering photon; and photon operating phototube. And does the idea of 'probability amplitude' make difficulty for us when the phototube operates a digital display? When a photon from the display enters the eye? When an impulse from the retina passes along the optic nerve? No, and again, no. Then why not also use probability amplitude to describe the state of the consciousness of the observer? In brief, does it make sense to 'include the observer in the wave function'? And if not, why not?

No question in physics is the takeoff point for a greater variety of considerations. They range from 'Schrödinger's (1935) cat' and 'Wigner's (1963) friend' to Everett's 'relative state' or 'many-universes' description of quantum mechanics [Everett (1957, 1973); DeWitt and Graham (1973); Cooper and Van Vechten (1969)] and from the difficulties pointed out by Bohr (1933) for 'measuring the state of the consciousness' to the nature of consciousness itself [see for example Ornstein (1973)]; and from 'non-separability' [d'Espagnat (1971a, 1971b); Bell (1975)] and the Einstein-Podolsky-Rosen (1935) paradox [see especially Bohr (1949, 1963)] to the question whether irreversibility is not an essential feature of any observing device [Bohr (1963), pp. 24–25].

Here we turn to another side, the cosmological side, of this much-considered question and ask, how can one possibly escape from including the observer in the wave function when any proper account of the dynamics of geometry would of necessity seem to have to include everything in its bookkeeping? The electromagnetic field responds only to charge, but the geometry of space responds in principle to all that bears mass-energy, from field to particle, from the inorganic to the organic, and from memory device to in-flight information. Along these lines to make the case as strong as possible for including everything, including the observer, in the wave function is the purpose of the first part of this report – in order that as much insight as possible can be gained in the last part by criticizing this case.

3. THE HAMILTON-JACOBI FORMULATION OF THE DYNAMICS OF GEOMETRY

What one means by the 'dynamics of space', and what its methods are, show nowhere more clearly than in the Hamilton-Jacobi formulation of Einstein's standard 1915 geometrodynamics, idealized to an 'empty' world. One is at liberty to include electromagnetic fields and other fields and particles as well, but only by adding complications that obscure what the dynamics of space is all about. A wave that carries energy from place to place can be illustrated as well by the gravitational field as by any other field; and so too can a black hole, and even an uncollapsed semistable concentration of energy or geon [Wheeler (1955)] that holds itself together by its own gravitational attraction in what looks from outside like any other mass. Limit attention to empty space: that is the one idealization introduced to simplify the discussion; the other: limit attention to general relativity in its Hamilton-Jacobi formulation. That version carries all the compulsion of any other classical statement of geometrodynamics, whether as field equation [Einstein (1915)], as variation principle [Hilbert (1915)], or as Hamiltonian system [Arnowitt et al. (1962)]. At the same time the HJ formulation makes the leap from the classical to the quantum description as short as possible, without actually leading into quantum gravity itself, with its unsolved questions of factor-ordering and of renormalization. [For a review see for example DeWitt (1967a, b, c) and the book edited by Isham et al. (1975.]

The HJ formalism, like every other classical formalism, has not one gene that is not of quantum theoretic origin. This one sees not least from the example of a particle of mass m and energy E moving on the x-axis under the influence of the potential $V(x)$. The HJ function

$$(1) \qquad S = S(x, t) = -Et + \int^x [2m(E - V(x)]^{\frac{1}{2}} \, dx,$$

solves the equation of propagation of wave-phase (dispersion relation; eikonal equation; HJ equation),

$$\begin{pmatrix} \text{rate of change} \\ \text{of dynamic} \\ \text{phase with time} \end{pmatrix} = \begin{pmatrix} \text{specified} \\ \text{function} \\ \text{of} \end{pmatrix} \begin{pmatrix} \text{rate of change} \\ \text{of dynamic} \\ \text{phase with position} \end{pmatrix},$$

or

$$\begin{pmatrix} \text{circular} \\ \text{frequency} \end{pmatrix} = \begin{pmatrix} \text{specified} \\ \text{function of} \end{pmatrix} \begin{pmatrix} \text{circular} \\ \text{wave number} \end{pmatrix},$$

or

$$-\partial S/\partial t = H(x, \partial S/\partial x),$$

or

$$(2) \qquad -\partial S/\partial t = (1/2m)(\partial S/\partial x)^2 + V(x).$$

The solution is spread all over (x, t) space. It gives not the slightest hint of anything like a classical world line. No more trace of localization does one see in the quantum wave function for a state of monochromatic energy,

$$\psi_E(x, t) \simeq \begin{pmatrix} \text{slowly varying} \\ \text{amplitude factor} \end{pmatrix} \begin{pmatrix} \text{rapidly varying} \\ \text{phase factor} \end{pmatrix}$$

$$(3) \qquad\qquad = A \exp(i/\hbar) S_E(x, t).$$

However, now superpose wave functions of slightly different energies,

(4) $\psi(x, t) = \psi_E(x, t) + \psi_{E+\Delta E}(x, t) + \cdots$.

The resulting wave packet is localized in the neighborhood of a point x. That point of constructive interference changes its position with time, t. The set of all such pairs of values (x, t) defines a world line. That world line is given by solving for x as a function of t, or t as a function of x, or both as functions $[x = x(\alpha), t = t(\alpha)]$ of some parameter α, in the condition of constructive interference,

$$S_E(x, t) = S_{E+\Delta E}(x, t)$$

or

(5) $\partial S_E(x, t)/\partial E = 0$.

This condition of interference reproduces immediately the familiar standard prediction for the motion,

$$0 = = \partial S/\partial E = -t + m \int dx[2m(E - V)]^{-\frac{1}{2}}$$

or

(6) $t = \displaystyle\int \dfrac{dx}{\left[\begin{array}{c}\text{velocity of particle of} \\ \text{energy } E \text{ at point } x\end{array}\right]}$.

A similarly simple conception of the dynamics of geometry was delayed for decades by the misleading notion that the dynamic object is spacetime. It is not spacetime. It is space. Three-dimensional space changes its shape with time. The history of that change is 4-dimensional spacetime; but the momentary configuration of space itself is 3-dimensional. Misunderstand this situation, formalistically try to apply Hamiltonian methods to the impossible subject of the dynamics of spacetime, and end up with answers like the identity that $0 = 0$!

The history of the particle is a world line. Make a 'cut' through that world line and arrive at a momentary configuration of that particle, the point (x, t).

The history of space is a spacetime, $^{(4)}\mathscr{G}$. Make a space-like slice through that spacetime and arrive at a momentary configuration of space, the 3-geometry $^{(3)}\mathcal{G}$.

How many parameters does it take to describe the shape of a potato? A 3-geometry needs more; it has one more dimension. Various possibilities for parametrizing a 3-geometry have been discussed in the literature [see for example Wheeler (1970)], but as simple as any is an approximate treatment in terms of a 'skeletonization' of the geometry [Regge (1961)] – and even simpler when it is postulated that the 3-geometry is closed. [For a discussion of the conflicting evidence for and against closure, including references to the recent literature, see for example Wheeler (1975).] A network of points at roughly equal separations divides up the space into tetrahedrons. The space inside each tetrahedron is idealized as Euclidean flatness. Thus the two faces that meet at the edge AB of a given tetrahedron are separated by a dihedral angle that is determined completely by the six edge lengths of that tetrahedron. The same is true for all the other tetrahedrons that hinge on the edge AB. The dihedral angles of this 'AB' set of tetrahedrons would sum to 2π if the 3-geometry were locally flat. The deficit from 2π, or 'angle of rattle', and the length and direction of AB, determine a kind of δ-function contribution to the curvature; similarly for other edges throughout the skeleton geometry. If there are 98 edges altogether in the framework, then the 3-geometry is fully specified, along with all details about its curvature, by giving the 98 edge lengths L_1, $L_2, \ldots, L_{98}$. Equivalently, one can give a single point in a Euclidean space ('truncated superspace') of 98 dimensions. The projections of that point on the 98 coordinate axes give the 98 members $L_1, L_2, \ldots, L_{98}$ and thus the momentary configuration of the 3-geometry.

Move this representative point a little in the truncated superspace, see the dimensions of the many tetrahedra change a little, and watch the 3-geometry alter everywhere in curvature and shape and change overall in volume. Starting from this model, one can imagine going to the mathematical limit of an infinitely fine skeletonization or otherwise representing a 3-geometry in all its detail by a single point in an ∞-dimensional superspace, [Riemann (1953 reprint); Wheeler (1970); Bers (1970); Fischer (1970)].

Superspace, $\mathcal{S}$, is the arena in which the dynamics of space unrolls, as (x, t) space is the arena in which the dynamics of the particle unrolls. The classical history of the particle is a world line, the set of the configurations (x, t) distinguished by the words, 'Yes, encountered', from the much larger 'No, not encountered in this history' set of (x, t)-values. The classical history of space is a spacetime, $^{(4)}\mathscr{G}$; but a spacetime appears in the context of superspace as a set of 3-geometries. Each such $^{(3)}\mathscr{G}$ is distinguished by this, that 'Yes, it is a $^{(3)}\mathscr{G}$ obtainable as a spacelike slice (wiggly or non-wiggly does not matter) through the given $^{(4)}\mathscr{G}$'. The generic $^{(3)}\mathscr{G}$ cannot be obtained by any spacelike slicing whatever through the given $^{(4)}\mathscr{G}$. These 'No, not obtainable as a spacelike slice' $^{(3)}\mathscr{G}$'s are infinitely more numerous than the 'Yes' $^{(3)}\mathscr{G}$'s. However, the 'Yes' $^{(3)}\mathscr{G}$'s are far too numerous to be accommodated on a 1-dimensional curve through $\mathscr{S}$. Instead, they fill out a manifold with one-third the dimensionality of $\mathscr{S}$ itself. This manifold is appropriately described as a 'leaf of history' extending through superspace.

What if one were to direct one's attention to quite another $^{(4)}\mathscr{G}$, quite another history of space evolving in time? Then slicing it would yield quite another set of $^{(3)}\mathscr{G}$'s, quite another leaf of history running through superspace, quite another classification of the totality of conceivable $^{(3)}\mathscr{G}$'s into 'Yes' $^{(3)}\mathscr{G}$'s and 'No' $^{(3)}\mathscr{G}$'s.

Returning to the original $^{(4)}\mathscr{G}$, consider a given slice $^{(3)}\mathscr{G}$ through it, and envisage this slice represented as a single point on the original leaf of history through superspace. Now recall the 'many-fingered' character of time in general relativity, a character strikingly recognized already in special relativity by Tomonaga (1946) in his 'bubble time' formulation of quantum electrodynamics. The observers spread out over the given spacelike hypersurface can report, not only what they see 'now', but also what they see a little time forward in the future. Moreover, the magnitude of this little time can vary a little from one point to another. Or, to adopt a jargon, one can 'push time forward' by different amounts in different places. This advance is characterized by one free function of three variables; or, more briefly, by 'one degree of freedom per space point'.

There are 'two other degrees of freedom per space point' in a $^{(3)}\mathscr{G}$. They are representable by two other functions of three variables. They represent 'gravitational-wave degrees of freedom'. Normally an alteration of this type carries attention from the given $^{(3)}\mathscr{G}$ to a nearby $^{(3)}\mathscr{G}$ that does not lie on the original leaf of history. Loosely stated, there are twice as many ways to move off the leaf as to stay on it.

This is the sense in which the 'leaf of history' spanned by the 'Yes' $^{(3)}\mathscr{G}$'s has one-third the dimensionality of the enveloping superspace.

The typical wave function $\psi(^{(3)}\mathscr{G})$ and the typical HJ- or 'phase'-function $S(^{(3)}\mathscr{G})$ are defined all over superspace. Localization on a leaf of history comes about only through constructive interference of many such individual waves, each characterized by parameters slightly different from those of the others. This is how the dynamics of space appears in the superspace description of geometrodynamics [Wheeler (1964); Misner *et al.* (1973), chapters 21 and 43].

Gerlach (1969) has shown that the evolution of geometry with time as calculated via constructive interference in this way satisfies all the demands of Einstein's standard field equation. Moreover, one knows [Kuchař (1974); Teitelboim (1973)] that any other HJ formulation of general relativity is equivalent to the superspace formulation, much as the description of a particle in terms of a conjugate momentum,

(probability amplitude) = (function of momentum)

is equivalent to a description in terms of the original coordinate itself,

(probability amplitude) = (function of position).

In this sense one can say that the probability amplitude is only then fully specified when the entire configuration of space, $^{(3)}\mathscr{G}$, or some geometric quantity conjugate to it in whole or in part, is fully specified. In brief, classical general relativity cannot hide the quantum sources of everything that goes on. Know the entire 3-geometry (or its conjugate), it says [by saying $S = S(^{(3)}\mathscr{G})$], or be deprived of what it takes for a complete quantum description of what is happening.

4. CAN ONE KNOW THE ENTIRE 3-GEOMETRY?

It is not necessary to know everything in order to know something. It is enough in classical Einstein geometrodynamics to know the 3- (or intrinsic) geometry and the associated time rate of change (or extrinsic) geometry for a finite region of space in order to be able to predict the future development of the geometry in some finite region for some finite reach of proper time. Moreover, one could claim that the ambitions of physics should be restricted to this limited kind of prediction. Available information, it could be argued, does not suffice for more. Up to today, $\sim 10 \times 10^9$ yr after the big bang, many parts of the universe have not yet been heard from. They cannot help sending us electromagnetic waves in the time to come. More relevantly, if for simplicity of discussion we continue to restrict attention to the extreme idealization of a purely geometrodynamic universe, no ir- regularly agitated faraway region can fail to agitate geometry here. However, it is enough for us to plant 'scouts', or detectors of gravitational radiation, out to a distance of a light year (or $100\,\ell$ yr) if we want to detect these effects. Then we can forecast the geometry in the neighborhood of an event E that is located here in space and located in time a year (or a century) into the future. Why isn't that kind of prediction a modest enough goal for physics?

At first sight such a 1-year (or 100-yr) prediction, far from being too modest compared to the possibilities, could be argued to be too ambitious. First, are we not too ambitious in our use of the word when we speak of a 'prediction'? The various waves from far away that wash in over our most remote scouts right now will not reach 'here' until the event E. But no processing by the scouts of the details of those waves on this 'right-now'-initial-value hypersurface, no sig- nals from these scouts, no warning about the geometry-to-come, can proceed faster than the speed of light. The last essential datum has no possibility to arrive at E ahead of the reality. Does anything built on such last-minute information deserve the name 'prediction'?

Second, are we not too ambitious in the coverage of space that we pretend to be able to achieve? Rather than information on the 3-geometry and its time rate of change on the initial-value hypersur- face out to the perimeter of scouting, σ_{scouted}, do we not more reasonably count on getting and using data in a still more restricted region of the hypersurface, σ_{precinct}? Then any 'prediction', last-minute in character as it is, is also of necessity an incomplete prediction.

5. CORRELATIONS RATHER THAN PREDICTIONS

In brief, the information available at E about what will happen at E is (1) in part last-minute and (2) ordinarily incomplete. For both reasons we speak better in most contexts of a 'correlation' than a 'prediction'. This is the sense in which Wigner states that "quantum mechanics can be ... reformulated in terms of the projection operator of the successive measurements.... [This] reformulation of the equations of quantum mechanics, eliminating explicit reference to the equations of motion and to state vectors, corresponds to a conceptual reformulation.... According to [it], the function of quantum mechanics is to give statistical correlations between the outcomes of successive observations." [Wigner (1973)]

To accept 'correlations' and thus in most circumstances to forego 'predictions' is realistic. It is the way to get on with the business of physics. Moreover, this version and vision of what physics is all about lays out a rich field for investigation: How are the relevant correlations to be defined? How does a correlation depend on the separation in space and in time of the two precincts in question? How are multiprecinct correlations related to 2-precinct correlations? What is the nature of the correspondence-principle transition from the quantum correlation function to the classical correlation function? What uncertainty relations characterize the quantum correlation functions? What procedures in principle offer themselves to verify the measurability of field quantities in two precincts right up to the limit of precision permitted by the appropriate uncertainty relation? For the case of electrodynamics some of these questions are treated, and treated wonderfully penetratingly and comprehensively, in a pair of papers unsurpassed for depth anywhere in theoretical physics [Bohr and Rosenfeld (1933, 1950)]. For the measurability of the dynamics of geometry the beginning of a beginning has been made [see for example Wigner and Saleckar (1958); Marzke and Wheeler (1964); DeWitt (1964); Ehlers et al. (1972); and Misner et al. (1973, p. 72)].

The ideal this work strives at, though not yet achieved, is clear: to describe in detail, even if idealized detail, the kind of equipment one would use and the kind of procedures one would employ, in order to measure an appropriate component of the curvature tensor in one precinct or pair of components in two precincts up to the precision indicated as possible by the standard type of field-theoretic commutation relations.

Not all colleagues agree that this field-theoretic ideal is attainable by any kind of experimental arrangement whatsoever, however idealized. Nevertheless, no detailed reasoning seems ever to have been offered in print to support this negative position. Moreover the history of the measurement problem in electrodynamics [Rosenfeld

(1955)] warns one how easy it is to devise what at first sight looks like a maximally effective measurement procedure [Landau and Peierls (1931)] which however does not and cannot come up to the precision of measurement demanded for the verification of the' theory. In quantum electrodynamics only faith in the field-theoretic predictions provided a lash powerful enough to stimulate the imagination to devise up-to-par measuring procedures; and one can well believe that the same history will mark quantum gravity.

To someone looking in from outside the subject this important ongoing discussion about whether one can achieve the predicted quantum limit of measurability might seem strange. Is it not, he could say, a little like a condemned prisoner debating between two forms of execution while all the time the question of overwhelming importance is, can he escape execution? 'Can predictability escape execution?' is to this onlooker the overriding issue. Or is it always to be true, signed, sealed and ratified by quantum mechanics, that the full data necessary for the prediction of an event never will be available and never can be available until the instant of the event? Should one not ask whether it fails before asking how it fails? It has often been said that 'predictability is the essence of science'. How then can one even talk of doing science when he admits the possibility that predictability may have to go by the board?

It is not necessary to return to the days of the Renaissance and astrology to take seriously the ideal of predictability. "[Jacob Burkhardt (1860)] finds it instructive – *lehrreich* – to see the hold of astrology on the Renaissance mind; neither education nor enlightenment, he insists, could do anything against this delusion – *Wahn* – because it was supported by the authority of the ancients and satisfied passionate fantasies and the fervent wish to know and determine the future [Gay (1974)]." The doctrine of predictability received a far sharper formulation and won a far more central position in the age of reason as evidenced not least in the famous statement of Laplace (1814): "Given for one instant an intelligence which could comprehend all the forces by which nature is animated and the respective situation of the beings who compose it – an intelligence sufficiently vast to submit these data to analysis – it would embrace in the same formula the movements of the greatest bodies of the universe and those of the lightest atom; for it, nothing would be uncertain and the future like the past would be present to its eyes."

No more impressive mathematical spelling out did that doctrine or any other doctrine ever receive than what one sees in the classical mechanics, the special relativity, the Maxwell electrodynamics and the Einstein geometrodynamics of the last century and this century. Moreover no outlook out of physics has ever won more currency in the other sciences than this classical concept of deterministic time evolution, with every step totally determined by the steps before it and the first step comprising the imagined complete specification of the initial value data. However today we have learned that we can know at most half of the needed initial data. Quantum mechanics incontestably rules out measuring more. It does not matter that there are islands of investigation where the concept of deterministic evolution in time still proves useful. It does not matter that in physics itself determinism is the right line of analysis for macroscopic events. For microscopic phenomena there is no escape from the lesson of quantum mechanics. One could imagine predicting the future position and velocity of a particle if one knew the present position and velocity; but no idea could be more misguided because no device whatsoever can acquire the needful information. Installation of equipment to measure the initial momentum automatically excludes all possibility to insert at the same time and in the same region the equipment that might measure its position; and conversely. Likewise equipment to measure the three components of the electric field over a region automatically excludes installation of the equipment that would measure the three components of the magnetic field in the same region at the same time, and conversely. Similarly, theory tells us, any attempt to measure the details of the intrinsic 3-geometry of a certain spacelike hypersurface automatically prevents us from measuring the full details of the extrinsic curvature of that same spacelike hypersurface; and conversely. Therefore enough information is available to predict in detail neither the world line of the particle nor the state of the electromagnetic field nor the 4-geometry that is the time history of the 3-geometry. It does not matter that one can construct in each case a wave packet: a wave packet 'fuzzed out' about the classical world line of the particle through spacetime; a wave packet diffusely enveloping the classical history of the electromagnetic field in spacetime; a wave packet spread out about the classical infinitely thin 'leaf of history' of 3-geometry in superspace. It does not matter that for many considerations the spread can be neglected and that to a good approximation one can use the words 'world line', 'deterministic Maxwell electrodynamics' and 'spacetime'. In principle every single one of these determinisms is contradictory to, and denied by, quantum mechanics. Predictability perishes.

In arriving at these familiar results, one or another version of the formalism of quantum mechanics has proven and undoubtedly always will prove indispensable; for example, the Schrödinger equation for the particle, the quantum mechanics of infinite systems of oscillators for the electromagnetic field; and propagation of a wave in superspace for the case of dynamics of geometry.

In all three cases the deterministic evolution of the wave function in its dynamic arena has sometimes been viewed as encouraging the hope that one can retain the idea of 'predictability' in one way or another. Nowhere is that hope expressed in more extreme form than in the 'relative state' or 'many-universes' formulation of quantum mechanics of Everett [Everett (1957) and Everett's fuller exposition in the book edited by DeWitt and Graham (1973); also other references reproduced in the same book]. In this treatment the 'memory coordinates of the observer' are included in the wave function along with the coordinates of the system under observation. Moreover, "apart from Everett's concept of relative states, no self-consistent system of ideas is at hand to explain what one shall mean by quantizing a closed system like the universe of general relativity [Wheeler (1957)]." In addition when one puts the classical dynamics of a closed universe into Hamilton-Jacobi form, and when one further recognizes that the Hamilton-Jacobi function $S(^3\mathscr{G})$ is the phase of the wave function and therefore contains information about the wave function itself, one comes directly to the quantum in the equation,

(7) $$\nabla^2_{(^{(3)}\mathscr{G})}\psi + {}^{(3)}R\psi = 0$$

for the dynamics of a closed universe. No room is left to put in the observer 'outside the system'. If he is to appear at all, he and his memory have to appear as particles and fields other than geometry; in other words, as additional variables on which the wave function has to depend. This proposed way of description, however, departs radically from anything to which one would be willing to give the name 'predictability' in any everyday sense of that word. Also the wave function is used in a sense quite different from that in which one uses it in any normal context. In particular great difficulties would seem to arise in giving any well defined meaning to the term 'state of the memory of the observer'. Thus as Bohr (1933) always emphasized, any attempt to "push the analysis of the mechanism of living organisms [i.e. the consciousness as ultimate observing device] as far as that of atomic phenomena ... would doubtless kill [the] animal" [and thus wipe out that very consciousness]. This central difficulty of the relative state or many-universes formulation has been emphasized especially in more recent times by Wigner (1971, 1973) and by d'Espagnat (1971b).

6. THE ROLE OF THE OBSERVER

What then in the light of the best present thinking does a knowledge of the wave function 'tell' one? It tells what one can observe. It tells the probability of a result when one makes the observation. It tells nothing, and can tell nothing, about the 'state of consciousness' of an observer.

The 'consciousness of the observer' is outside the wave function. An observation is only then an observation when it is recorded in the consciousness [Wigner (1974)]. An observation is only then an observation when one observer can tell another the result of the observation 'in plain language' [Bohr (1962)].

That the 'observer' should have a special place in the scheme of quantum mechanics has often been contested. Bohr himself argued at one point (1963, pp. 24–25) that an "irreversible amplification process" is all it takes to "complete" a measurement, only to stress on other occasions that an observation is only complete when there is an observer (1933 and elsewhere). It takes only an act of nuclear fission to illustrate the essential considerations. For all normal circumstances that process is irreversible when one considers the variety of fission products and the number of secondary neutrons and gamma rays that come out. However, in principle one can imagine 'mirrors' set up to give sufficient time delays and sufficient accuracy of return of the outgoing particles and radiations to bring everything back together again 'in phase' and undo the fission process. In this sense the act of fission does not meet the ultimate test of a quantum mechanical measurement process in the sense that it isn't 'indelible' [Moldauer (1968) and Belinfante (1975)]. Fission, and by extension the pulse of a Geiger counter and the blackening of a silver halide crystal, are only then guaranteed to be indelible in the relevant sense when the act has registered in the consciousness of the observer.

In summary it does not appear reasonable to 'include the observer in the wave function'. On this account it can hardly be judged legalistically acceptable to speak of a 'wave function for the universe and all it contains'. Still less has any slightest indication ever been found of any possibility to reach beyond the laws of physics itself to a 'universal wave function' [Wheeler (1967)], derivable from considerations of simplicity or any other way that would allow any return whatsoever to predictability from a new direction. Rather, the wave function will continue in the future as in the past to be a formalistic device, not to 'predict' in the nineteenth century sense of prediction, but to forecast correlations between observations. As for the size of the region covered by those observations nothing is said. It can be as small as an atom. It can have the extension of a superconductor. It can have to do with correlations between photons from a distant star. Or it can have to do with the diagnosis of geometry over a very extensive region of the universe. If in the process of making such large scale forecasts it is convenient to go to the limiting idealization of a 'wave function for the whole universe' – and it is often more convenient to deal with the whole than a part – one will not imply by that procedure that one has the slightest intention of trying to give a meaning to the impossible concept of 'a probability amplitude for this, that, or the other state of the whole universe'. Instead one will be understood merely to have adopted a convenient formalism for calculating correlations that do have a meaning.

In such calculations the concepts of '3-geometry', 'superspace', 'dynamic phase, $S(^{(3)}\mathcal{G})$' and 'probability amplitude, $\psi(^{(3)}\mathcal{G})$', forever occupy central positions, as the corresponding concepts of 'configuration', 'dynamic arena', 'Hamilton-Jacobi function' and 'wave function' forever hold key places in any other domain of classical or quantum dynamics. Superspace is here to stay.

Princeton University and
University of Washington

NOTE

* Battelle Memorial visiting professor during the period of preparation of this report for publication. Permanent address since 1 September 1976: Physics Department, University of Texas, Austin, Texas 78712.

16   J. A. WHEELER

BIBLIOGRAPHY

Arnowitt, R., Deser, S., and Misner, C. W., 1962, 'The Dynamics of General Relativity', in Witten (1962), pp. 227–265.

Belinfante, F. J., 1975, *Measurements and Time Reversal in Objective Quantum Theory*, Pergamon, Oxford, p. 39.

Bell, J. S., 1975, 'The Theory of Local Beables', preprint TH2053 from CERN, Geneva.

Bers, L., 1970, 'Universal Teichmüller Space', pp. 65–84 in Gilbert and Newton (1970); see also reference 51 on p. 376 of Wheeler (1970).

Bohr, H., 1933, 'Light and Life', *Nature* **131**, 421–423 and 457–459; reprinted in Bohr (1958). See p. 9 of the latter for the quoted words.

Bohr, N., 1949, 'Discussion with Einstein on Epistemological Problems in Atomic Physics', pp. 201–241 in Schilpp (1949); reprinted in Bohr (1958), pp. 32–66.

Bohr, N., 1958, *Atomic Physics and Human Knowledge*, Wiley, New York [cited under Bohr (1933) and Bohr (1949)].

Bohr, N., 1962, 'Light and Life Revisited', p. 3 and pp. 23–29 in Bohr, N., 1963; see especially pp. 24–25.

Bohr, N., 1963, *Essays 1958–1962 on Atomic Physics and Human Knowledge*, Wiley-Interscience, New York [cited under Bohr (1962)].

Bohr, N. and L. Rosenfeld, 1933, 'Zur Frage der Messbarkeit der elektromagnetischen Feldgrössen', *Kgl. Danske Videnskab. Selskab* **12**, No. 8.

Bohr, N. and Rosenfeld, L., 1950, 'Field and Charge Measurements in Quantum Electrodynamics', *Phys. Rev.* **78**, 794–798.

Burckhardt, J., 1860, *Die Kultur der Renaissance in Italien*, Leipzig; authorized English translation from the 15th ed. by S. G. C. Middlemore, 1929, *The Civilization of the Renaissance in Italy*, Harper, New York.

Carmeli, M., Fickler, S. I., and Witten, L., 1970, *Relativity*, Plenum, New York; cited under Fischer (1970).

Chiu, H.-Y. and Hoffman, W. F. (eds.), 1964, *Gravitation and Relativity*, W. A. Benjamin, New York.

Cooper, L. N. and Van Vechten, D., 1969, 'On the Interpretation of Measurement within the Quantum Theory', *Am. J. Phys.* **37**, 1212–1220; reprinted in DeWitt and Graham (eds.), 1969.

d'Espagnat, B. (ed.), 1971, *Foundations of Quantum Mechanics*, Academic, New York. [Cited under Wigner (1971).]

DeWitt, B., 1964, Dynamical Theory of Groups and Fields', pp. 587–822 in DeWitt and DeWitt (eds.) (1964); see esp. pp. 598–614 on measurement and observables.

DeWitt, B. S., 1967a, 'Quantum Theory of Gravity. I.', *Phys. Rev.* **160**, 1113–1148.

DeWitt, B. S., 1967b, 'Quantum Theory of Gravity. II; the Manifestly Covariant Theory', *Phys. Rev.* **162**, 1195–1239.

DeWitt, B. S., 1967c, 'Quantum Theory of Gravity III; Applications of the Covariant Theory', *Phys. Rev.* **162**, 1239–1256.

DeWitt, B. S. and Graham, N. (eds.), 1973, *The Many-Worlds Interpretation of Quantum Mechanics*, Princeton University Press, Princeton, New Jersey [cited under Everett (1957 and 1973), Wheeler (1957), Cooper and Van Vechten (1969).

DeWitt, C. and DeWitt, B., 1964, *Relativity, Groups and Topology*, Gordon and Breach, New York [cited under DeWitt (1964) and Wheeler (1964)].

Ehlers, J., Pirani, F. A. E., and Schild, A., 1972, 'The Geometry of Free Fall and Light Propagation', in O'Raifeartaigh (1972), pp. 63–84.

Einstein, A., 1915, 'Die Feldgleichungen der Gravitation', *Preuss. Akad. Wiss. Berlin, Sitzber.*, 844–847.

Einstein, A., Podolsky, B., and Rosen, N., 1935, 'Can Quantum-Mechanical Description of Physical Reality be Considered Complete', *Phys. Rev.* **47**, 777–780.

Everett, H. III, 1957, '"Relative State" Formulation of Quantum Mechanics', *Rev. Mod. Phys.* **29**, 454–462; reprinted in DeWitt and Graham (eds.), 1973.

Everett, H. III, 1973, 'The Theory of the Universal Wave Function', pp. 3–140 in DeWitt and Graham (eds.), 1973.

Fischer, A. E., 1970, 'The Theory of Superspace', in Carmeli, Fickler, and Witten (1970).

Gay, P., 1974, *Style in History*, Basic Books, New York.

Gerlach, U., 1969, 'Derivation of the Ten Einstein Equations from the Semiclassical Approximation to Classical Geometrodynamics', *Phys. Rev.* **177**, 1929–1941.

Gilbert, R. P. and Newton, R. G., 1970, *Analytic Methods in Mathematical Physics*, Gordon and Breach, New York; cited under Bers (1970); Wheeler (1970).

Gold, T. (ed.), 1967, *The Nature of Time*, Cornell University Press, Ithaca, New York [cited under Wheeler (1967)].

Hilbert, D., 1915, 'Die Grundlagen der Physik', *König. Gesell. d. Wiss. Göttingen, Nachr., Math.-Phys. Kl.* 395–407.

Hooker, C. A. (ed.), 1973, *Contemporary Research in the Foundations and Philosophy of Quantum Theory*, Reidel, Dordrecht, Holland. [Cited under Wigner (1973).]

Isham, C., Penrose, R., and Sciama, D., 1975, *Quantum Gravity*, Clarendon Press, Oxford.

Kuchař, K., 1974, 'Geometrodynamics Regained: a Lagrangian Approach', *J. Math. Phys.* **15**, 708–715.

Landau, L. and Peierls, R., 1931, 'Erweiterung des Unbestimmtheitsprinzips für die relativistische Quantentheorie', *Zeits. f. Physik* **69**, 56–69.

Marzke, R. F. and Wheeler, J. A., 1964, 'Gravitation as Geometry. I. The Geometry of Spacetime and the Geometrodynamical Standard Meter', pp. 40–64 in Chiu and Hoffman (eds.), 1964.

Misner, C. W., Thorne, K. S., and Wheeler, J. A., 1973, *Gravitation*, Freeman, San Francisco; cited under Regge (1961).

Moldauer, P. A., 1968, 'Measurement, Memory, and Classical States', *Bull. Am. Phys. Soc.* **13** (No. 2), 180, Abstract ED2.

O'Raifeartaigh, L. (ed.), 1972, *General Relativity*, *Papers in Honor of J. L. Synge*, Oxford University Press, London.

Ornstein, R. E. (ed.), 1973, *The Nature of Human Consciousness: A Book of Readings*, W. H. Freeman, San Francisco, California, 1973.

Pauli, W. (ed.), 1955, *Niels Bohr and the Development of Physics*, McGraw-Hill, New York [cited under Rosenfeld (1955)].

Regge, T., 1961, 'General Relativity without Coordinates', *Nuovo Cimento* **19**, 558–571; see also Misner, Thorne, and Wheeler (1973), chapter 42.

Riemann, B., 1953, 'Theorie der Abel'schen Funktionen', pp. 88–142, and 'Vorlesungen über die allgemeine Theorie der Integrale algebraischer Differentialen', Supplement I, as reprinted in his *Gesammelte Mathematische Werke*, H. Weber (ed.), 2nd ed., Dover Publications, New York.

Rosenfeld, L., 1955, 'On Quantum Electrodynamics', pp. 70–95 in Pauli, W. (ed.), 1955.

Schilpp, P. A. (ed.), 1949, *Albert Einstein: Philosopher-Scientist*, Library of Living Philosophers, Evanston, Illinois and subsequent paperback reprints elsewhere; cited under Bohr (1949).

Schrödinger, E., 1935, 'Die gegenwartige Situation in der Quantenmechanik', *Naturwissenschaften* **23**, 807–812.

Seeger, R. J. and Cohen, R. S., 1974, *Philosophical Foundations of Science*, Reidel, Dordrecht; cited under Wigner (1974).

Shaviv, G. and Rosen, J. (eds.), 1975, *General Relativity and Gravitation: Proceedings of the Seventh International Conference (GR7), Tel-Aviv University June 23–28, 1974*, Wiley, New York [cited under Wheeler (1975)].

Teitelboim, C., 1973, 'How Commutators of Constraints Reflect the Spacetime Structure', *Ann. Phys.* **79**, 542–557.

Tomonaga, S., 1946, 'On a Relativistically Invariant Formulation of the Quantum Theory of Wave Fields', *Prog. Theor. Phys.* **1**, 27–42.

Wheeler, J. A., 1955, 'Geons', *Phys. Rev.* **97**, 511–536.

Wheeler, J. A., 1962, 'The Universe in the Light of General Relativity', *The Monist* **47**, 40–76; see also Wheeler (1964), pp. 517 and 520.

Wheeler, J. A., 1964, 'Geometrodynamics and the Issue of the Final State', in DeWitt and DeWitt (1964), pp. 312–522; see sections on superspace.

Wheeler, J. A., 1967, 'Three-dimensional Geometry as Carrier of Information About Time', pp. 90–107, and especially pp. 103–107, in Gold (1967).

Wheeler, J. A., 1970, 'Superspace', pp. 335–378 in Gilbert and Newton (1970).

Wheeler, J. A., 1975, 'Conference Summary', pp. 299–344 in Shaviv and Rosen, 1975; see esp. pp. 223–224.

Wigner, E. P., 1963, 'The Problem of Measurement', *Am. J. Phys.* **31**, 6–15.

Wigner, E. P., 1971, 'The Subject of Our Discussions', pp. 1–19 in d'Espagnat, B. (ed), 1971.

Wigner, E. P., 1973, 'Epistemological Perspective on Quantum Theory', pp. 369–385 in Hooker, C. A. (ed.) (1973). See especially pp. 375–376 on correlations and pp. 382–383 on the idea of a "wave function for the whole universe".

Wigner, E. P., 1974, 'Physics and the Explanation of Life', pp. 119–132 in Seeger and Cohen (1974).

Wigner, E. P. and Saleckar, H., 1958, 'Quantum Limitations of the Measurement of Space-Time Distances', *Phys. Rev.* **109**, 511–577.

Witten, L., (ed.), 1962, *Gravitation: An Introduction to Current Research*, Wiley, New York; cited under Arnowitt *et al.*, 1962.

A. FRENKEL*

ON THE POSSIBLE CONNECTIONS BETWEEN
QUANTUM MECHANICS AND GRAVITATION

ABSTRACT. Since those memorable days when Louis de Broglie had the brilliant idea of associating a wave with the electron, the problems of the interpretation of this wave and its relations with observed phenomena have been and remain an intriguing source of puzzles for us. Not only are we far from agreement on the answers, we often disagree even more about the sort of questions to ask.

In this brief report I would like to present a possible way to formulate the main problem in the theory of measurement and a particular attempt to solve it. I wish to make clear that the ideas and the work carried out on the basis of these ideas belong to Professor F. Károlyházi, who is working at the Fötvös University in Budapest. The results are contained in his second thesis defended in 1972, published only in Hungarian[1].

A brief note on the general ideas has been published[2]. The thesis is much more complete and contains important results which are not even mentioned in the note of 1966. With the authorization of F. Károlyházi I have prepared this article making use of his thesis. I had to select material, of course, in a way which was inevitably not impartial. Nevertheless I hope that what I am going to say is not in contradiction with the author's view.

I. TWO OPEN PROBLEMS

1. The reduction of the wave function is not described by Schrödinger's equation. Moreover, this equation – and, more generally, the formalism of quantum mechanics – does not even tell us when, and in what circumstances the reduction must be done. In other words, we do not have theoretical criteria to distinguish a microsystem from a macrosystem. The theory of quantum mechanics teaches us that de Broglie's idea is valid for any system of mass M, i.e. that we have to associate with this mass its Compton wave length,

$$(1) \qquad L = \frac{h}{Mc},$$

which, together with the velocity v of the system gives the de Broglie wave length,

$$(2) \qquad \lambda = L \sqrt{\frac{c^2}{v^2} - 1} \underset{v \ll c}{\approx} L\frac{c}{v},$$

J. Leite Lopes and M. Paty (eds.), Quantum Mechanics, a Half Century Later, 19–38. All Rights Reserved
Copyright © 1977 by D. Reidel Publishing Company, Dordrecht-Holland

where λ is the wave length associated with the quantum mechanical wave behaviour of the centre of mass of the system. According to quantum theory the wave behaviour exists for systems with any mass. On the other hand, we know that the quantum mechanical wave function ceases to be an adequate representative for the description of massive systems, since it does not account for their observed classical behaviour. Empirically, an atom of the table is a micro-system, the table itself is a macrosystem. A smooth transition between these two should exist, and a step forward would be made if the theory could indicate where it is.

It is worth noting that to interpret the results given by Schrödinger's equation we have to admit the existence of the so-called 'measuring apparatuses' to which this equation does not apply. This means in fact that we admit that the validity of the superposition principle is limited. If wanted, we may push this limit very far, the observer's consciousness being the ultimate 'measuring apparatus' which escapes the rules of quantum mechanics. In this case such pieces of measuring apparatuses as a Stern-Gerlach magnet, a plate of nuclear emulsion as well as all the rest of the system except the consciousness of the aforementioned observer, should be described by Schrödinger's equation. However, as we have learned from J. von Neumann and, in another context, from N. Bohr, we can put the limit nearer to the microworld and describe all these measuring apparatuses as classical systems. Such a situation, although logically admissible, is nevertheless unsatisfactory because on the one hand the theory does not tell us how far the limit can be pushed towards the microworld. On the other hand, it means that to interpret quantum mechanics we need systems – at least one system – which escapes the laws of quantum mechanics. A possibility to overcome the latter difficulty is contained in the theory of B. Everett with superposed Universes. The solution proposed by Károlyházi also overcomes it in a way which seems more simple and natural. Also, his solution provides us with a theoretical criterion to distinguish a microsystem from a macrosystem, which is not the case in Everett's theory.

2. Another open question which at first sight has nothing to do with the former one is the connection between quantum mechanics and the theory of general relativity of Einstein. Let us note, first of all, that this classical (i.e. non-quantum) theory also pretends to have an

unlimited validity. Any system is submitted to the law of gravity, and the mass distribution in the Universe determines (up to symmetry transformations) a unique, sharp metric $g_{\mu\nu}(x_\rho)$ at each world point x_ρ. Note also that in its turn this theory associates a length – the Schwarzschild radius –

$$(3) \qquad r = 2\frac{G}{c^2}M,$$

with any system of mass M, G being the gravitational constant. The sphere of radius r limits that region around the centre of mass from which (light) signals cannot come out.

It is remarkable that these two theories – quantum mechanics and general relativity – may claim an unlimited validity for the superposition principle and for the sharp metric of spacetime only if they ignore each other. Indeed, according to quantum mechanics,

$$(4) \qquad \Delta x\, \Delta v \geq \frac{h}{M}.$$

Equation (4) shows that the distribution of the centre of mass of any system with mass $M < \infty$ is subject to fluctuations. These fluctuations imply fluctuations of the metric (determined, as we know, by the distribution of the masses), in contradiction with the idea of a sharp, unique value of $g_{\mu\nu}$ at each point x_ρ. On the other hand, quantum mechanics is discussed, as a rule, with the Minkowski metric,

$$(5) \qquad -g_{00} = g_{11} = g_{22} = g_{33} = -1; \qquad g_{\mu\nu} = 0 \quad \text{if } \mu \neq \nu,$$

for all x_ρ. If the fluctuations of the metric are taken into account, it is conceivable that for two terms of a linear superposition a spread of order π in their relative phase may arise, and this can be interpreted as a destruction of the coherence between these terms. In this way a quantitative theoretical criterion may arise for the limits of validity of the superposition principle.

The idea that the fluctuations of the metric may destroy the coherence is not new. It has been put forward, among others, by R. P. Feynman. However, this is an idea which is difficult to exploit. Indeed, the length which first emerges in connection with this idea is the universal length of Planck Λ:

(6) $\dfrac{1}{2} rL = \dfrac{hG}{c^3} \equiv \Lambda^2 \approx (10^{-33} \, \text{cm})^2,$

and it is desperately small. It corresponds to a mass,

(7) $m_\Lambda = \dfrac{h}{\Lambda c} \approx 10^{-5} \, \text{g},$

which, for normal densities ($\approx 1 \, \text{g cm}^{-3}$) and normal temperatures (≈ 300 K) empirically is a macroobject, and thus Λ can hardly be as simply connected with the transition region between micro and macro as suggested by (7).

Professor Wheeler explains in this volume how Planck's length could play an important role at the time of the Creation, just after the big-bang. Following Károlyházi we shall now realize that this length may play a fundamental role in quieter epochs, too. Namely, it does enter the equations which allow one to distinguish the microsystems from the macrosystems, but in a more subtle way than in (7). For instance, we shall find that for a solid the transition region may be characterized by a mass

(8) $M^{\text{tr}} \approx 10^{-14} \, \text{g}.$

This corresponds to a colloidal grain, which consists of $\approx 10^{10}$ atoms, a result more plausible and more encouraging than (7).

II. SCHRÖDINGER'S EQUATION ON A SPACETIME WITH SMEARED METRIC

Let us now look at the main steps of the construction of the proposed theory. Some parts of the derivation are given in the appendix.

1. One shows (see the appendix) that the uncertainty relation (4) together with (3) leads to an uncertainty ΔT for the precision with which a time interval T can be measured:

(9) $(\Delta T)^3 \geqslant \dfrac{\Lambda^2}{c^2} T = \dfrac{hG}{c^5} T.$

Note first of all how small this bound is. For $T = 1$ s, $\Delta T \geqslant 10^{-29}$ s!

On the other hand this bound, in contradistinction to that for Δx in (4), is absolute, with no alternative: ΔT cannot be reduced by increasing the uncertainty in the value of some complementary quantity. Note also that (9) is a non-linear relation between ΔT and T and that the coefficient consists only of the universal constants characterizing the two theories involved.

Károlyházi argues that (9) indicates that when we build our theories using a spacetime with sharp metric $g_{\mu\nu}(x_\rho)$ we push an abstraction coming from good old classical physics too far. Indeed, the idea of a sharp metric implies that there is nothing preventing, in principle, the measurement of T with complete precision. (9) shows that such a bound does exist and accordingly we should go beyond the idea of spacetime with a sharp metric.

2. To make the relation (9) an inherent part of quantum mechanics one constructs a model of spacetime with a smeared metric which is consistent with (9). Namely, a family $(g_{\mu\nu}(x_\rho))_\beta$ of metrics, each very close to the Minkowski metric (5) is introduced. β is a random variable which labels the members of the family. The (proper) time interval T between two world points $x_\rho^{(1)}$, $x_\rho^{(2)}$ is then defined to be the mean value of the time intervals T_β, calculated with their corresponding metrics $(g_{\mu\nu})_\beta$. Thus

$$(10) \qquad T = \bar{T}_\beta,$$

whereas the natural definition of ΔT is

$$(11) \qquad \Delta T = \sqrt{\overline{(T - T_\beta)^2}}.$$

Károlyházi has shown (see the appendix) that the family $(g_{\mu\nu})_\beta$ can be chosen in such a way that for T and ΔT given by (10) and (11) the relation (9) holds for any pair of points $x_\rho^{(1)}$, $x_\rho^{(2)}$ which lie on a world line corresponding to a motion with velocity $v \ll c$. It can be shown that this restriction implies that the theory in its present form cannot be applied to macrosystems whose relative velocity is relativistic. Although important from a conceptual point of view, this restriction has no bearing on the discussion of the main problems of the theory of measurement.

To avoid possible misunderstanding, I would like to stress that the metrics $(g_{\mu\nu})_\beta$ are classical (i.e. non-quantized) fields, and no physical significance should be attached to the individual members of the family $(g_{\mu\nu})_\beta$. Their only role is to provide us with a model of spacetime with smeared metric in agreement with (9). We explained why this model of spacetime should be nearer to the truth than a spacetime with sharp Minkowski metric. The lack of a unified theory of gravitation and quantum mechanics prevents one from doing much better at present. Another source of misgivings might be that the choice of the family $(g_{\mu\nu})_\beta$ leading to (9) is not unique. However, it turns out that the results are independent of the details of this choice.

3. The third step is to study the propagation of the quantum mechanical wave functions on our spacetime with a smeared metric. In quantum mechanics we are used to working almost exclusively with the Minkowski metric (5). Let us assign the value $\beta = 0$ of the set $\{\beta\}$ to this particular metric and denote correspondingly the Schrödinger wave function propagating on flat space by ψ_0. The Schrödinger equation for ψ_0 is then the usual one:

$$(12) \qquad i\hbar\frac{\partial}{\partial t}\psi_0(t) = H\psi_0(t).$$

Here and below we shall consider possibly complicated but always *isolated* systems; therefore H does not depend on time explicitly.

It is clear that ψ_0 does not adequately describe the propagation of the system on our spacetime with smeared metric. To arrive at such a description, we have to also consider all the state vectors ψ_β – one for each member of the family $(g_{\mu\nu})_\beta$. Since all the $(g_{\mu\nu})_\beta$'s are close to the Minkowski metric, we may write the Schrödinger equation for ψ_β in a non-relativistic approximation in the form

$$(13) \qquad i\hbar\frac{\partial}{\partial t}\psi_\beta(t) = (H + V_\beta)\psi_\beta(t).$$

In (13) H is the same energy operator as in (12) and V_β is the gravitational potential energy due to the deviation of the metric $(g_{\mu\nu})_\beta$ from the Minkowski metric. To be more specific, for a single particle of mass m (13) gives

$$(14) \qquad i\hbar\frac{\partial}{\partial t}\psi_\beta(\mathbf{x}, t) = \left(-\frac{\hbar^2}{2m}\Delta + c^2 m\gamma_\beta(\mathbf{x}, t)\right)\psi_\beta(\mathbf{x}, t).$$

The construction of the function

(15) $\qquad \gamma_\beta(\mathbf{x}, t) \equiv (g_{00}(\mathbf{x}, t))_\beta - 1 \equiv \dfrac{1}{mc^2} V_\beta(\mathbf{x}, t),$

is carried out in the appendix. The other components of $(g_{\mu\nu})_\beta$ do not contribute in the non-relativistic approximation. In (14) the spin has been neglected, but it could be incorporated without difficulty. For a system of N (not necessarily identical) particles we have

(16) $\qquad i\hbar \dfrac{\partial}{\partial t} \psi_\beta(X, t) = \Bigg[\displaystyle\sum_{i=1}^{N} -\dfrac{\hbar^2}{2m} \Delta_i + \sum_{\substack{i,k=1 \\ i \neq k}}^{N} V_{ik}$

$$+ c^2 \sum_{i=1}^{N} m_i \gamma_\beta(\mathbf{x}_i, t) \Bigg] \psi_\beta(X, t).$$

In (16) and below X denotes a point $[\mathbf{x}_1, \ldots, \mathbf{x}_N]$ of the configuration space. The 'particles' may be the nucleons and the electrons of an atom, or the atoms of a macroscopic body, or the atoms or the molecules of that body etc., depending on the nature of the system and on the approximation used. In (16) V_{ik} is the interaction energy between the particles. It is very important to note that in spite of the fact that it is often difficult or impossible to find the solutions of (16) even when

(15') $\qquad V_\beta(X, t) \equiv c^2 \displaystyle\sum_{i=1}^{N} m_i \gamma_\beta(\mathbf{x}_i, t) \equiv \sum_{i=1}^{N} V_\beta(\mathbf{x}_i, t),$

is put equal to zero (i.e. for the usual case), it is possible to study how the V_β's perturb these poorly known solutions and to arrive at remarkable quantitative conclusions in many interesting cases. The general method can be outlined as follows:

Let the system be represented at some 'initial' moment $t = 0$ on all the metrics $(g_{\mu\nu})_\beta$ by the same[1] wave function

(17) $\qquad \psi_\beta(X, 0) = \psi_0(X, 0).$

At some later time $t > 0$ the ψ_β's will no longer be equal to each other. Instead, because of the smallness of the V_β's, we shall find to a good approximation

(18) $\qquad \psi_\beta(X, t) \approx \psi_0(X, t) \, e^{i\phi_\beta(X, t)},$

where

(19) $\phi_\beta(X, t) = -\dfrac{1}{\hbar} \displaystyle\int\limits_0^t V_\beta(X, t')\, dt'.$

The spread

(20) $\overline{[(\phi_\beta(X^{(1)}, t) - \phi_\beta(X^{(2)}, t))^2]^{\frac{1}{2}}},$

in the relative phase between two points $X^{(1)}$, $X^{(2)}$ of the configuration space can be calculated as a function of these points and of time. It is immediately clear that the spread vanishes for $t \to 0$ and also for $X^{(2)} \to X^{(1)}$. On the other hand, with increasing separation in the configuration space and with increasing time the spread in the relative phase may reach the value π. These features lead to a natural interpretation of the physical state represented by the set $\{\psi_\beta\}$ which gives rise to the difference between microsystems and macrosystems.

III. MICROSYSTEMS AND MACROSYSTEMS

1. Let us first consider the propagation of a stable elementary particle of mass m. The configuration space now consists of the single coordinate variable x (we neglect the spin) which in this particular case coincides with the centre of mass coordinate q. The gravitational potentials $V_\beta(q, t)$ are known (see (15)) and therefore the spread δ in the relative phase between two points $q^{(1)}$ and $q^{(2)}$ can easily be calculated. The result turns out to depend only on the distance

(21) $a \equiv |q^{(1)} - q^{(2)}|,$

between the two points:

(20′) $\overline{[(\phi_\beta(q^{(1)}, t) - \phi_\beta(q^{(2)}, t))^2]^{\frac{1}{2}}} = \delta(a, t),$

where

(22) $\phi_\beta(q, t) = -\dfrac{1}{h} \displaystyle\int\limits_0^t V_\beta(q, t')\, dt'.$

The function $\delta(a, t)$ for some fixed value of a is represented on the figure

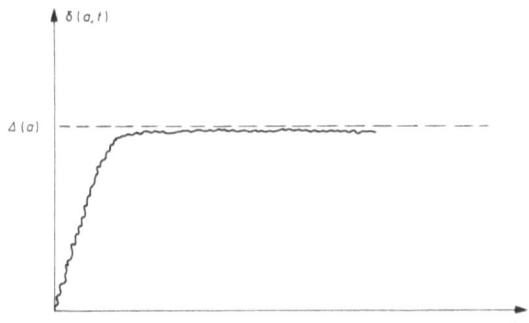

δ starts from zero at $t = 0$, reaches very rapidly – with small oscillations – an asymptotic value $\Delta(a)$ and remains practically equal to it afterwards. The function $\Delta(a)$ is zero for $a = 0$ and increases with a. For two points separated by a distance such that

(23) $\Delta(a) \ll \pi$,

the spread $\delta(a, t)$ will remain much less than π for all t's, and the relative phases will practically be independent of β. On the other hand for some sufficiently large value a_c of a we shall arrive at

(24) $\Delta(a_c) \approx \pi$.

The relative phase between two points separated by a distance equal to or larger than a_c will have a spread of order π. Let us call a domain of dimension a_c in configuration space a 'coherence cell' and divide the configuration space into such cells.[2] Suppose that at $t = 0$ the particle is confined to a single cell (i.e. $\psi_0(\mathbf{x}, t) = \psi_\beta(\mathbf{x}, 0) = 0$ outside the cell). If the particle remained in this single cell for all time, then the coherence of the state represented by $\{\psi_\beta\}$ would be preserved for all time. The system would then behave as if it had a single coherent wave function $\psi_0(\mathbf{x}, t)$, each ψ_β being just one of the rays belonging to ψ_0. However, it is well known that the Schrödinger equation for an isolated system leads to spread of the wave function. Namely, after a time

(25) $\tau_c \approx \dfrac{ma_c^2}{h}$,

the wave function $\psi_0(\mathbf{x}, t)$ will spread to a region of $\approx 2a_c$. Then for points belonging to different cells the set $\{\psi_\beta\}$ will no more behave as a single coherent wave function, and the relative phases between these points will change violently with β. Following Károlyházi we shall interpret this situation as an indication that the system is now in a state which is no longer represented adequately by a single ray, and therefore we have to make the reduction of the wave function in such a way that the coherent parts be again confined to single cells, the weight being given by the absolute value of the amplitude in each cell.

Thus the propagation of an isolated system can be described in terms of 'expansion-reduction cycles', the period of a cycle being τ_c. In the cycle both the causal (Schrödinger equation) and the stochastic (reduction) part of the propagation are included. The stochastic part now has its physical cause – the spread in the metric of spacetime – and therefore it is conceived as the description of a physical process which takes place independently of any observer. The theory in its present form can only indicate why and when the reduction takes place, the mathematical equations which would describe the reduction itself are lacking. Therefore we are forced to make the reduction by hand, just as in the orthodox theory. However we now know *when* the reduction takes place, and we shall shortly see that this provides us with a mathematical criterion for distinguishing a microsystem from a macrosystem.

As indicated above, a_c can be calculated, and for an elementary particle of mass m the result is

$$(26) \qquad a_c \approx L \left(\frac{L}{\Lambda} \right)^2,$$

where L is the Compton wave length of the particle. For an electron (26) and (25) give

$$(27) \qquad a_c \approx 10^{35}\,\text{cm}, \qquad \tau_c \approx 10^{70}\,\text{s}.$$

These values exceed astronomical scales. Thus there is no chance for an isolated electron to expand into a region corresponding to two cells and then undergo a reduction, and there is no chance for us to observe this phenomenon. We arrive at the result that for an isolated electron only the causal behaviour is important and accordingly it finds its adequate mathematical representation in a set $\{\psi_\beta\}$ which is always equivalent to a single coherent wave function $\psi_0(\mathbf{x}, t)$. The situation is the same for any elementary particle, e.g. for a proton

$a_c \approx 10^{25}$ cm, $\tau_c \approx 10^{53}$ s. For obvious reasons we shall call such systems 'microsystems'.

2. We now come to the problem of massive systems. First of all I wish to stress that 'massive' also means that the system inevitably has many degrees of freedom, and this fact plays an important role in the theory we discuss.

When we wish to describe the motion of a massive system as a whole, we usually represent it by a single degree of freedom – its centre of mass coordinate – and we associate the whole mass with this degree of freedom. However this cannot be done without further ado in the theory of Károlyházi. In particular, his formula (26) cannot be extended to massive systems just by saying that a_c now refers to the centre of mass coordinate and L is the Compton wave length corresponding to the total mass. We shall see presently why this is so.

Let us discuss the case of a solid ball of mass M, radius R, volume Ω and density d. We have of course

$$(28) \qquad M = d\Omega = d\frac{4\pi}{3}R^3.$$

We consider a ball of ordinary density ($d \approx 1\,\mathrm{g\,cm^{-3}}$) and we suppose that the ball is isolated in spite of the fact that it may be difficult or even impossible to isolate it if M is large. However, we wish to show that the spread in the metric of spacetime is *sufficient* to explain how microbehaviour goes over into macrobehaviour with increasing M. The interactions of the ball with its surroundings (thermal contact, etc.) may and generally do modify the details,[3] but are not the cause of the macrobehaviour.

Let the ball consist of N particles with masses and coordinates x_i. In a solid it is useful to put

$$(29) \qquad x_i = x_i^0 + r_i,$$

where r_i is the deviation of the ith particle from its equilibrium position x_i^0. If the state of the ball corresponds to normal (i.e. not too high) temperature, then in $V_\beta(X, t)$ we may write as a good approximation x_i^0 instead of x_i:

$$(30) \qquad V_\beta(X, t) \approx c^2 \sum_{i=1}^{N} m_i \gamma_\beta(x_i^0, t) \approx c^2 d \int_\Omega dx' \gamma_\beta(x', t).$$

(Obviously, the integral is an excellent approximation for the sum.) The integral in (30) can be calculated and it depends of course on the position of the centre of the ball which coincides with its centre of mass coordinate $\mathbf{q}$, and on its radius R. This shows why the system as a whole cannot be simply represented by its mass concentrated in the centre of mass. Indeed, it turns out that if d increases by a factor k, the integral will not simply diminish by that factor. The dependence on R is essential.

Thus we find that for our ball V_β is a known function of $\mathbf{q}$, R, d, t:

$$(31) \qquad V_\beta(X, t) \approx V_\beta(\mathbf{q}, R, d, t).$$

One can now carry out the calculation of the coherence length for the centre of mass coordinate $\mathbf{q}$ of the ball along the same lines as in the case of the elementary particles. Of course we have to take V_β from (31). One finds

$$(32) \qquad a_c \approx L\left(\frac{L}{\Lambda}\right)^2 \qquad \text{for} \quad a_c \gtrsim R,$$

$$(33) \qquad a_c \approx L\left(\frac{R}{\Lambda}\right)^{\frac{2}{3}} \qquad \text{for} \quad a_c \lesssim R,$$

where as usual

$$(34) \qquad L = \frac{h}{Mc}.$$

We obtain here quite a remarkable result. We learn from these equations that when the coherence length a_c for the centre of mass of the ball is larger than its radius R, the radius drops out from the formula for a_c and moreover the formula coincides[4] with the one for the elementary particles! On the other hand, when $a_c \lesssim R$, then the formula for a_c is different from the elementary particle one, and, in particular, it depends on R – a genuine macroscopic parameter. Therefore we come to a natural division of our balls into three categories:

(a) $a_c \gg R$ microbehaviour dominates,
(b) $a_c \approx R$ transition region,
(c) $a_c \ll R$ macrobehaviour dominates.

We shall presently have a closer look at case (a), but let us first see how massive a ball corresponding to the transition region is. For $a_c \approx R$ (32) and (33) coincide and give

(35) $a_c \approx R \approx \dfrac{L^3}{\Lambda^2}.$

Using (34) and (28) we find for normal densities

(36) $M^{tr} \approx 10^{-14}\,\mathrm{g}, \qquad R^{tr} \approx a_c^{tr} \approx 10^{-5}\,\mathrm{cm}.$

This is the size of a colloidal grain. The period of its cycle is

(37) $\tau_c^{tr} \approx 10^3\,\mathrm{s}.$

Thus the theory affirms that if a colloidal grain could be isolated for a few hours, then its wave function would not only expand according to the Schrödinger equation, but would also undergo several reductions without any measurement being accomplished. This result constitutes a departure from the predictions of the orthodox theory, and in principle this departure should be observable.

Indeed, according to orthodox theory the wave function of an isolated system suffers dispersion only as a result of Schrödinger's equation. According to the proposed theory the spread is due to both Schrödinger's equation and to the non sharp structure of spacetime, the latter leading to repeated reductions of the wave function. As a result, the spread in the position of the centre of mass is larger than in the orthodox theory. The tiny energy needed for this supplementary spread comes from the gravitational potentials V_β which act formally as time dependent external fields. The reaction of the system on the V_β's (i.e. its contribution to the spread of the metric) is not taken into account in the theory. This would make it necessary to go beyond the phenomenological model of spacetime, a task which probably cannot be accomplished at present.

Unfortunately, the isolation of a colloidal grain during such a long time seems to be technically impossible. However we shall see that an experiment to detect the departure from the orthodox theory can probably be realized.

An important qualification concerning Equations (32)–(33) for a_c should be made. They are relevant for the behaviour of the system as a whole only if the spread of the wave function due to the other degrees of freedom is slower than the spread due to the centre of mass coordinate. Indeed, it would happen that the r_i's in (29) are negligible when calculating V_β, but not negligible when studying the spread of the wave function according to Schrödinger's equation. As a matter of fact, for a ball larger than 10 cm the expansion due to the inner degrees of freedom becomes faster than the expansion due to the centre of mass coordinate. This shows that in each particular case one has to analyse the expansion-reduction cycle not only with respect to the centre of mass coordinate, but also with respect to the other degrees of freedom. As an example we mention that for superconductors the spread associated with the electrons carrying the superconducting behaviour is slow, this part of the system remains confined to a single cell for very long times. On the other hand, the spread associated with the degrees of freedom of the 'raw material' is fast. The system is a macrosystem in this respect. The interaction between these two parts is weak, which makes their coexistence possible. All this is in full agreement with the observed phenomena.

3. Let us now look at the behaviour of a ball with $R \approx 1$ cm. This is certainly a macrosystem, both empirically and according to our criterion. How would such a ball propagate if it were isolated?

We easily find that in this case

(38) $a_c \approx 10^{-16}$ cm, $\tau_c \approx 10^{-4}$ s.

This means first of all that two states for which the positions of the centre of mass of the ball are separated by a distance larger than 10^{-16} cm lose their coherence. The coherent parts of the state $\{\psi_\beta\}$ describing the centre of mass of the ball are confined to single coherence cells. This means that the centre of mass is confined to one of these cells. (Of course if we do not know in which one it is, we shall describe it by a mixture of coherent states $\{\psi_\beta\}$.) 10^{-4} s later, owing to Schrödinger's equation two cells will be occupied by the wave functions $\{\psi_\beta\}$. At this moment the coherence gets destroyed and the reduction must be carried out in order to account for it. The ball is now either in one, or in the other of these cells, and so on. Due to the repeated cycles the centre of mass of the ball accomplishes an 'anomalous Brownian motion'. The mean velocity v_c associated with this motion is

(39) $v_c \approx \dfrac{a_c}{\tau_c} \approx c\dfrac{L}{a_c}.$

$v_c \approx 10^{-12}$ cm s^{-1} for the ball in question. Thus in the proposed theory macrobehaviour does not exactly mean classical behaviour. The ball does not move along a perfectly classical trajectory. However, the deviation from it amounts only to tiny zig-zags characterized by (38) and (39) and the deviation diminishes with increasing mass.

It is important to realize that the aforementioned anomaly (with respect to orthodox theory) is inherent in the behaviour of any system. v_c turns out to be largest for a colloidal grain, i.e. in the transition region ($v_c \approx 10^{-8}$ cm s^{-1}). For microsystems it again decreases and for an electron it is $\approx 10^{-35}$ cm s^{-1}. This reflects the fact that the colloidal grain shows an appreciable departure both from the purely classical and from the purely quantum behaviour. Its wave function expands according to the Schrödinger equation during a few hours, but it undergoes reduction at the end of these periods. As we go towards the microsystems, the quantum behaviour takes over, and for an electron the classical behaviour does not manifest itself in practice owing to the huge values of a_c and τ_c. Thus classical and quantum behaviour appear in the theory of Károlyházi as the limits of the macro and microbehaviour respectively, with a smooth transition between them.

IV. AN EXPERIMENTAL PROPOSAL

It is hopeless to try to isolate the aforementioned ball with $R \approx 1$ cm. However, it is possible to suspend it in a gas with the help of a thin thread. It can be shown that in this case the mean displacement of the centre of mass from its equilibrium position due to the anomalous Brownian vibration will exceed in a few hours the amplitude of the normal Brownian vibration by a factor of 10^4. The magnitude of this signal is equal to that of the expected noises coming from the fluctuations of the gravitational field of the Earth, from the possible distortions in the mechanism of the suspension, etc. The technical details are under study. Unfortunately there seems to be no easier way for testing the difference between the two theories.

V. CONCLUSION AND OUTLOOK

In the proposed theory the old problem of the reduction of the wave
function is treated from a new standpoint. By combining Heisenberg's
uncertainty relation with gravitation, an absolute quantitative limi-
tation on the sharpness of the structure of spacetime is derived. Then
the effect of this uncertainty is incorporated into the quantum
mechanical equation of motion. The propagation of any isolated
system is partly causal (Schrödinger's equation, spread of the wave
function), partly stochastic (reduction of the non-coherent parts of
the wave function due to the spread in the metric). The interplay of
these aspects leads to the introduction of the concepts of the coherence cell and of the expansion-reduction cycle characterized by
the coherence length a_c and the period τ_c respectively. For elementary particles as well as for systems for which $a_c \gg R$, the causal
aspect dominates, the coherence length and the period of the cycle
are enormous ($a_c \approx 10^{35}$ cm, $\tau_c \approx 10^{70}$ s for the electron) and they do
not have enough space and time to manifest their macroscopic
properties. On the other hand, for systems with $a_c \ll R$ a_c and τ_c are
small (10^{-16} cm and 10^{-4} s for a solid ball of 1 g) and their behaviour is
nearly classical. The superposition principle has a limited validity for
these systems even if they are isolated, and coherence survives only
as long as the spread in the position of the centre of mass of the
system is confined to a single coherence cell. If the system is
sufficiently massive the reduction will take place very often, leading
to the characteristic macrobehaviour.

When a microsystem interacts with a macrosystem in such a way
that the interaction induces a sufficiently large change in the mass
distribution of the latter (when its 'zero' state is in a different
coherence cell than its 'interaction took place' state), the reduction
takes place for the macrosystem, and the microsystem coupled to it
also undergoes a reduction. This is the measurement process in the
proposed theory. The need for an ab initio classical measuring apparatus or for a conscious observer disappears.

As I emphasized above, the theory of Károlyházi is not a substitute
for a unified theory of quantum mechanics and gravitation. In my
opinion he succeeded in expressing an essential feature of this future
theory – the spread in the metric of spacetime – in a simple and
fruitful way, and in constructing on this basis a theory which ac-

counts for the transition between microbehaviour and macrobe-haviour. Undoubtedly, a unified theory will go deeper and further and will modify the theory presented here, but I believe that the most important results will reappear.

APPENDIX

1. *The Derivation of Relation* (9)

Let M be the mass of the 'hand' of a clock. To measure a time interval T between to moments t_0 and t_1 we must read the position of the hand at these moments. At the moment t_0 we have

$$(A1) \qquad \Delta x_0 \, \Delta v_0 \geqslant \frac{h}{M},$$

where Δv_0 is the spread in the velocity of the centre of mass of the hand. Therefore the uncertainty Δx_1 in the position at the moment t_1 will be

$$(A2) \qquad \Delta x_1 \approx \Delta v_0 T \geqslant \frac{h}{\Delta x_0 M} T.$$

The resulting uncertainty in T will therefore be

$$(A3) \qquad \Delta T \approx \frac{\max (\Delta x_0, \Delta x_1)}{v},$$

where v is the mean velocity of the hand. (A2) shows that for fixed M and T ΔT will be minimal when

$$(A4) \qquad \Delta x_0 \approx \Delta x_1 \equiv \Delta x.$$

Therefore from (A2) and (A3) we find

$$(A5) \qquad (\Delta T)^2 \geqslant \left(\frac{\Delta x}{v}\right)^2 \geqslant \frac{h}{v^2 M} T \geqslant \frac{h}{c^2 M} T.$$

Let us now take into account that the hand of the clock cannot be confined to a region smaller than its Schwarzschild radius. Therefore we have

$$(A6) \qquad \Delta x \geqslant r \approx \frac{MG}{c^2}.$$

(A5) and (A6) give

(A7) $$(\Delta T)^3 \geqslant \frac{hG}{c^5} T \equiv \frac{\Lambda^2}{c^2} T,$$

which is our relation (9).

2. The Construction of the Family $(g_{\mu\nu})_\beta$

Let

(A8) $(g_{\mu\nu})_\beta = (g_{\mu\nu}) = 0$ when $\mu \neq 0$ and/or $\nu \neq 0$

(A9) $\begin{cases} (g_{00})_0 = 1 \\ (g_{00})_\beta = 1 + \gamma_\beta(x, t) \quad \text{for} \quad \beta \neq 0. \end{cases}$

We shall construct the γ_β's by writing down their Fourier series expansions:

(A10) $$\gamma_\beta(x, t) = \frac{1}{\sqrt{L^3}} \sum_k [c_\beta(k)\, e^{i(kx - \omega t)} + c_\beta^*(k)\, e^{-i(kx - \omega t)}].$$

In (A10) L is the length of the edge of an arbitrarily chosen large box and has nothing to do with the Compton wave length also denoted by L in the text.

As usual, we have

(A11) $k = \dfrac{2\pi}{L}\mathbf{n}$ with n_x, n_y, n_z integers,

and

(A12) $\omega = c|\mathbf{k}| = ck,$

because

(A13) $\Box \gamma_\beta(x, t) = 0.$

Let us now choose an integer

$$N_k \geqslant 2,$$

for each $\mathbf{k}$ and introduce the random variables $b(\mathbf{k})$ which may take N_k values

(A14) $b(\mathbf{k}) = \dfrac{2\pi}{N_k} \cdot [0, 1, \ldots, N_k - 1].$

A particular choice for the values of these variables will give a particular set $c_\beta(\mathbf{k})$ through the definition

(A15) $c_\beta(\mathbf{k}) = \Lambda^{\frac{2}{3}} k^{-\frac{5}{6}} e^{ib(\mathbf{k})}$.

Thus each set of the $b(\mathbf{k})$'s corresponds to a (non zero) value of the set $\{\beta\}$. We give equal weight to each value of β. Then it is easy to see that

(A16) $\overline{(c_\beta(\mathbf{k}))^m} = 0$ for $m = 1, 2, \ldots,$

(A17) $\overline{|c_\beta(\mathbf{k})|^2} = \Lambda^{\frac{4}{3}} k^{-\frac{5}{3}}$.

Consider now two world points A, B lying on a world line $x^\mu(t)$. The proper time interval s_β between them is given by

(A18) $s_\beta = \dfrac{1}{c} \displaystyle\int_A^B \sqrt{g_{\mu\nu} \dfrac{dx^\mu}{dt} \dfrac{dx^\nu}{dt}} \, dt; \quad x^0 \equiv ct.$

For simplicity we shall carry out the calculation only for world lines along which $\mathbf{v} \equiv d\mathbf{x}/dt = 0$. In this case for $A(\mathbf{x}, t_1)$, $B(\mathbf{x}, t_2)$ we find

(A19) $s_\beta = \displaystyle\int_{t_1}^{t_2} \sqrt{1 + \gamma_\beta(\mathbf{x}, t)} \, dt \approx \int_{t_1}^{t_2} (1 + \tfrac{1}{2}\gamma_\beta(\mathbf{x}, t)) \, dt.$

(A16) shows that $\overline{\gamma_\beta(\mathbf{x}, t)} = 0$ and we find

(A20) $s \equiv s_\beta \approx t_2 - t_1.$

Furthermore (A16) gives

(A21) $(\Delta s)^2 \equiv \overline{(s - s_\beta)^2} = \dfrac{1}{(2\pi)^3} \displaystyle\int d\mathbf{k} \dfrac{\overline{|c_\beta(\mathbf{k})|^2}}{\omega^2} [1 - \cos \omega (t_2 - t_1)],$

and (A17) leads to

(A22) $(\Delta s)^2 \approx \left(\dfrac{\Lambda^2}{c^2} s \right)^{\frac{2}{3}},$

in agreement with (9). For world lines along which $v \ll c$ (A22) remains valid.

We see that only the averages (A16) and (A17) are important for the derivation of (A22). Similarly, they alone enter the calculation for the spread of the relative phase of the state $\{\psi_\beta\}$. Therefore any realization of the family $(g_{\mu\nu})_\beta$ which leads to (A16) and (A17) is acceptable.

Finally we note that when $k^{-1} \leqslant 10^{-13}$ cm we are in the terra incognita of the structure of the elementary particles. Therefore the theory should be independent of the details of the choice of the $c_\beta(\mathbf{k})$'s in that region. This is indeed the case. E.g. a cut-off $c_\beta(\mathbf{k}) = 0$ for $k^{-1} \leqslant 10^{-13}$ cm will not change the results.

Orsay, France

NOTES

* Permanent address: Central Research Institute for Physics, Budapest, Hungary.
[1] It would be more natural to start with slightly different ψ_β's on the different $(g_{\mu\nu})_\beta$'s. The result would be unchanged but the argument would be more complicated.
[2] The way in which this division should be carried out is thoroughly discussed in the thesis.
[3] A very interesting discussion of these problems is contained in the thesis.
[4] There are some unessential deviations between them, hidden in our $\approx$ sign, which always stands for equality in order of magnitude.

BIBLIOGRAPHY

[1] F. Károlyházi, *Magyar Fizikai Folyóirat* **12** (1974), 23.
[2] F. Károlyházi, *Nuovo Cimento* **52** (1966), 390.

J. M. JAUCH

THE QUANTUM PROBABILITY CALCULUS*

> At bottom, the theory of probability is only
> common sense reduced to calculation.
>
> Pierre Simon Laplace, 1812

> Probability is the most important concept in
> modern science, especially as nobody has the
> slightest notion what it means.
>
> Bertrand Russell, 1929

I. INTRODUCTION

Quantum mechanics has opened a vast sector of physics to probability calculus. In fact most of the physical interpretation of the formalism of quantum mechanics is expressed in terms of probability statements.[1]

There are of course large segments of classical physics, too, which are expressed in probabilistic terms. But there is an essential difference between the probabilistic statements of quantum physics and those of classical physics. The present article is devoted to the elucidation of this difference.

The probabilities which occur in classical physics are interpreted as being due to an incomplete specification of the systems under consideration, caused by the limitations of our knowledge of the detailed structure and development of these systems. Thus these probabilities should be interpreted as being of a *subjective* nature.

In quantum mechanics this interpretation of the probability statements has failed to yield any useful insight, because it has not been possible to define an infrastructure whose knowledge would yield an explanation for the occurrence of probabilities on the observational level. Although such theories with 'hidden variables' have been envisaged by many physicists,[2] no useful result has come from such attempts. I therefore take here the opposite point of view which holds that the

J. Leite Lopes and M. Paty (eds.), Quantum Mechanics, a Half Century Later, 39–62. All Rights Reserved
Copyright © 1977 by D. Reidel Publishing Company, Dordrecht-Holland

probabilities in quantum mechanics are of a fundamental nature deeply rooted in the objective structure of the real world. We may therefore call them *objective* probabilities.

It has been noted quite early that the probabilities in quantum theory have some peculiar properties, unrelated to anything previously encountered in classical probability theory. One way of exhibiting these anomalies is by studying joint probabilities for certain pairs of random variables, for instance, those corresponding to the quantum-mechanical position q and the canonically conjugate momentum p.[3] For this case it has been noted by Wigner (1932) already that no positive joint distribution exists.

Various interpretations have been given of this anomaly. I shall not review them critically here, but rather offer yet another one, which I believe corresponds better to the objective character of the quantum probability calculus than previous interpretations.

One point of departure is the observation that the Wigner anomaly for the joint distribution of noncompatible observables is an indication that the classical probability calculus is not applicable for quantal probabilities. It should therefore be replaced by another, more general calculus, which is specifically adapted to quantal systems. In this article I exhibit this calculus and give its mathematical axioms and the definitions of the basic concepts such as probability field, random variable, and expectation values.

Generalized probability calculi have been proposed before.[4] My proposal differs in several respects from previous work on this subject insofar as it is specifically motivated by and adapted to the axiomatic structure of quantum theory as it has been developed by the Geneva School[5] since 1960.

II. PROBABILITY CALCULUS AND PROBABILITY THEORY

The proposed modification of the probability calculus appears more natural if we distinguish between *probability calculus* and *probability theory*.[6] With *calculus* we denote the mathematical formalism devoid of any interpretation of this formalism. With *theory* we refer to the the application of this calculus to various situations involving the occurrences of observable phenomena.

The calculus is a branch of mathematics (in fact of measure theory) and presents no problems of interpretation. The theory on the other hand is beset with numerous difficulties which have been the object of much controversy.

It is remarkable that in none of these controversies was the calculus as such ever questioned and its definitive form as given by Kolmogorov[7] in 1933 has been the basis of all the work on mathematical statistics. The slight generalization of this calculus by Renyi[8] is not essentially different insofar as it removes the restriction of a normalized total probability and replaces it by the basic notion of *conditional* probability.

Little thought has been given to the question why this particular calculus should be so effective in predicting the probabilities of actually occurring events.

The logical situation that we are facing here may be illustrated by an analogy from another branch of mathematics. The discovery of geometry by the Greeks, and in particular its axiomatization by Euclid, led to the idea that the geometry of physical space was unique and absolute. The discovery of non-Euclidean geometries was at first thought to be of no relevance to the geometry of physical space. Only in the physics of the twentieth century, especially through the work of Hilbert and Einstein, did the idea break through that physical geometry is not Euclidean and can actually be determined objectively through physical observations.

Clearly geometry plays the role of the calculus and its interpretation in terms of physical phenomena. It is conceivable that the general theory of relativity could be expressed on the background of a Euclidean space, but in the light of present knowledge it would not be *natural* to do so.

In an analogous way, we contend, it would be possible to express quantum theory on the background of a classical probability calculus, but again, Wigner's work has clearly shown that it would not be natural either to do so.

So just as the geometry of space-time is determined by physical phenomena in the context of a natural theory, it is my belief that probability calculus is equally determined by certain phenomena in the context of quantum theory.

In order to place the new calculus in the proper perspective, I begin with a commentated review of the classical probability calculus.

III. THE CLASSICAL PROBABILITY CALCULUS

The classical calculus of probability is based on a few concepts which I shall introduce and comment briefly in this part. The concepts are: the measurable space, the probability measure, the random variables, the probability distribution function, and the expectation values.

1. *The Measurable Space*

The primary concept of probability calculus is 'the universe of basic events' which in the classical case are identified with a certain class $\mathscr{S}$ of subsets of a set Ω.

The set Ω may be completely arbitrary. Actually as we shall see this set plays in fact only a subsidiary role. What is important are the subsets of the class $\mathscr{S}$ which shall be called the measurable sets.

The class of subsets $\mathscr{S}$ is assumed to be a 'field'. This means it is closed with respect to the operations of the complement, countable unions, and intersections. Furthermore it contains ϕ, the null set, and consequently also Ω, the entire set.

Thus if $S \subset \mathscr{S}$ then the complementary set $S' \in \mathscr{S}$. If S_n $(n = 1, 2, \ldots)$ is a countable family of sets from $\mathscr{S}$ then

$$\bigcup_n S_n \in \mathscr{S} \quad \text{and} \quad \bigcap_n S_n \in \mathscr{S}.$$

2. *The Probability Measure*

On the field $\mathscr{S}$ is defined a positive-valued function

$$\mu : \mathscr{S} \to \mathbb{R}^+$$

with the properties

(i)　　　$\mu(\phi) = 0$;　　$\mu(\Omega) = 1$.

(ii)　　　For any pairwise disjoint sequence S_n $(n = 1, 2, \ldots)$ such that $S_i' \subset S_k$ for $i \neq k$

$$\mu(\bigcup_n S_n) = \sum_n \mu(S_n) \quad (\sigma\text{-additivity}).$$

This function is the probability measure on $\mathscr{S}$.

We shall refer to the triplet $(\Omega, \mathscr{S}, \mu)$ as the *probability space*. The interpretation of this calculus is that the sets $S \in \mathscr{S}$ denote the possible 'events' and the numbers $\mu(S)$ represent the 'probability' for the occurrence of these events.

3. Random Variables

Let $X:\Omega \to \mathbb{R}$ be a real-valued function $X(\omega)$, $\omega \in \Omega$. For any subset $\Delta \in \mathbb{R}$ we denote by

$$X^{-1}(\Delta) = \{\omega \mid X(\omega) \in \Delta\}$$

the *inverse image* of the set Δ under the function X.

A function f is said to be *measurable-B* or simply measurable if for every Borel set $\Delta \in \mathfrak{B}(\mathbb{R})$ the inverse image $X^{-1}(\Delta) \in \mathcal{S}$.

A real *random variable* is a real-valued measurable function on Ω.

It will be seen in the following that the essential property of a random variable, in fact the only property which is really used, is the correspondence which it establishes between Borel sets $\Delta \in \mathfrak{B}(\mathbb{R})$ and the measurable sets. In view of the proposed generalization it is useful to introduce a special notation for this correspondence. Thus we shall denote by $\xi : \mathfrak{B}(\mathbb{R}) \to \mathcal{S}$ the correspondence set up by the random variable $X(\omega)$ through

$$X^{-1}(\Delta) = \xi(\Delta),$$

and we shall call ξ also a random variable.

This correspondence has the following properties:

(i) $\xi(\phi) = \phi \in \mathcal{S}$; $\xi(\mathbb{R}) = \Omega$.

(ii)   If $\Delta_i \perp \Delta_k$ for $i \neq k$ then $\xi(\Delta_i) \perp \xi(\Delta_k)$

(disjoint sets are mapped into disjoint sets).

(iii) $\xi(\bigcup \Delta_n) = \bigcup_n \xi(\Delta_n)$

for any pairwise disjoint sequence Δ_n.

If X_0, X_1, and X_2 are random variables, i.e., measurable functions, then so are $X_1 + X_2$, $X_1 X_2$, X^{-1} (if it exists) and for any sequence X_n ($n = 1, 2, \ldots$) $\lim \sup X_n$, $\lim \inf X_n$, and $\lim X_n$ (if the limit exists).

4. The Distribution Function

Let X be a random variable and denote by

$$S_a \equiv \xi((-\infty, a])$$

then
$$F_\xi(a) = \mu(S_a)$$

is called the *distribution* function of the random variable ξ in the probability space $(\Omega, \mathscr{S}, \mu)$.

It has the following properties:

(i) $F_\xi(a)$ is a nondecreasing function, continuous from the right and it tends to the limit 0 as $a \to -\infty$.

(ii) $F_\xi(\infty) = 1$.

If $F_\xi(a) = 1$ is continuous and absolutely continuous then we may define a probability density $f_\xi(a) \geqslant$ by setting

$$\frac{\mathrm{d}F_\xi(a)}{\mathrm{d}a} = f_\xi(a).$$

The derivative exists everywhere.

5. *The Expectation Value*

Let ξ be a random variable, $F(a)$ its distribution function, then we define the *expectation* value by the integral (if it exists)

$$\langle \xi \rangle = \int_{-\infty}^{+\infty} a \, \mathrm{d}F(a) = E(X).$$

This is also called the *mean value* of ξ in $(\Omega, \mathscr{S}, \mu)$.

The notation is chosen deliberately in order to adumbrate the proposed generalization. The expression on the right-hand side is the classical one, while the left-hand side is used for the quantal one.

If ξ_1 and ξ_2 are two random variables represented by their measurable functions X_1 and X_2 we denote by $\xi_1 + \xi_2$ the random variable represented by $X_1 + X_2$. Similarly if ξ is represented by X then ξ^2 is represented by X^2.

With this notation we find for the *variance*

$$D^2(\xi) = \langle (\xi - \langle \xi \rangle)^2 \rangle = E((X - E(X))^2)$$

or

$$D^2(\xi) = \langle \xi^2 \rangle - \langle \xi \rangle^2 = E(X^2) - E(X)^2.$$

The notion of independent random variable is of great importance in probability calculus. We formulate it here also in a generalizable fashion first for sets.

Two sets A, $B \in \mathscr{S}$ are said to be independent with respect to the probability measure μ if

$$\mu(A \cap B) = \mu(A) \, \mu(B).$$

The notion can be generalized to n sets $A_1, A_2, \ldots, A_n \in \mathscr{S}$. They are independent if and only if for every $i_1, i_2, \ldots, i_m$ $(m \leqslant n)$

$$\mu(A_{i_1} \cap A_{i_2} \cap \cdots \cap A_{i_m}) = \mu(A_{i_1}) \, \mu(A_{i_2}) \ldots \mu(A_{i_m}).$$

The notion can be extended to random variables. The random variables ζ and η represented by the measurable functions X and Y are independent with respect to μ if for any pair A, $B \in \mathfrak{B}(\mathbb{R})$ of Borel sets on the real line

$$\mu(\xi(A) \cap \eta(B)) = \mu(\xi(A)) \, \mu(\eta(B)).$$

These are the essential concepts of the classical probability calculus.

IV. THE PROBABILITY CALCULUS IN CLASSICAL MECHANICS

For a classical mechanical system the probability space Ω is the classical phase Γ. The probability measure for a system with no restriction will be the Lebesgue measure on Γ. Liouville's theorem assures that this measure is invariant under the evolution of the system due to the classical equations of motion.

Actually in isolated systems it is not this measure which can be used since it is not normalizable to one. Isolated systems will be restricted to a surface of constant energy. This measure is called the microcanonical measure and it is only defined on the surfaces of constant energy. If the system is not isolated but kept at a constant temperature by thermal contact with a heat bath then it is the canonical measure which is appropriate.

Every state of the system defines a new kind of measure. In particular a 'pure' state is given by a measure concentrated in one point $\omega \in \Gamma$. We shall denote it by δ_ω. It is defined explicitly by

$$\delta_\omega(A) = \begin{cases} 1 & \text{for} \quad \omega \in A \\ 0 & \text{for} \quad \omega \notin A. \end{cases}$$

The distribution function $F(a)$ of a random variable ζ for a pure state

δ_ω is defined by

$$F(a) = \delta_\omega(\xi((-\infty, a])) = \begin{cases} 1 & \text{for} \quad \omega \in \xi((-\infty, a]) \\ 0 & \text{for} \quad \omega \notin \xi((-\infty, a]). \end{cases}$$

For such a state the expectation value of the random variable ξ is given by

$$\langle \xi \rangle = \int_{-\infty}^{+\infty} a \, dF(a) = a_0$$

where a_0 is the smallest value a for which $\xi((-\infty, a]) = 1$.

V. THE PROBABILITY CALCULUS IN QUANTUM MECHANICS

The preceding discussion of the probability calculus in classical mechanics serves the purpose of illustrating the need for generalizing this calculus if it is intended for application in quantum physics.

The first important observation is the absence of the phase space Γ in quantum mechanics. Hence it is necessary to develop a probability calculus without Kolmogorov's set Ω used for the definition of the measure space. At first sight this seems impossible since it would seem to make the definition of random variables impossible. However this is not so.

A careful examination of the classical probability calculus reveals that it could have been developed without ever mentioning the set Ω. The only place where this is not obvious is in the definition of random variables which we have defined as measurable functions $X(\omega)$ of $\omega \in \Omega$. However the subsequent use of these functions consisted merely in establishing σ-homomorphism $\xi: \mathfrak{B}(\mathbb{R}) \to \mathscr{S}$ through the formula

$$\xi(\Delta) = X^{-1}(\Delta) \in \mathscr{S} \quad \text{for all} \quad \Delta \in \mathfrak{B}(\mathbb{R}).$$

Hence the calculus can be reconstructed in its entirety without ever mentioning Ω if we define random variables by this homomorphism. Of course in this case the class $\mathscr{S}$ must no longer be considered as consisting of the subsets of a set. Instead we replace it by a set of elements for which union, intersection, and complement is defined, in short $\mathscr{S}$ is a lattice.

In the classical case the lattice $\mathscr{S}$ was of a special kind, called a Boolean lattice, which is characterized by the distributive law

$$A \cap (B \cup C) = (A \cap B) \cup (A \cap C)$$
$$A \cup (B \cap C) = (A \cup B) \cap (A \cup C). \tag{0}$$

Once we have freed ourselves from the special interpretation of the lattice $\mathscr{S}$ as subsets of a set there is no need for maintaining the distributive law.

The structure of the lattice of 'elementary events' – we shall call them propositions, or yes-no experiments – will have to be determined from experiment and this involves a physical interpretation of the operations $\cap$, $\cup$ and the complement. This has been done in the case of quantum mechanics, and the result is that the operations of union and intersection lead to a non-Boolean lattice. This is the essential feature of general quantum mechanics.

I should perhaps mention here for completeness that there have been attempts to represent the quantum-mechanical proposition system on a weaker structure, the partially ordered sets (or posets).[9] The reason is that it is not always possible to exhibit in an operational manner the meet $A \cap B$ of two elementary events. However, as shown in Jauch (1968),[10] there are situations where this is possible even for a noncompatible pair of propositions A, B. This is always the case if there exist two passive filters, which represent measurements of the first kind corresponding to these two propositions. There exists then a filter $A \cap B$ which is obtained as an infinite alternating sequence of filters A and B. This is the operational analogue of the well-known formula $E \cap F = S - \lim_{n \to \infty} (EF)^n$ for the meet of two not necessarily commuting projection operators E and F in Hilbert space.

We shall denote by $\mathscr{L}$ the lattice of elementary events (propositions) in quantal physics and by $a, b, c, \ldots$ the elements from $\mathscr{L}$.

We have a partial-order relation in $\mathscr{L}$ denoted by c, as well as the operations of join and meet $a \cup b$ and $a \cap b$. They define the greatest lower bound and the least upper bound of a and b.

The lattice of propositions is orthocomplemented. The orthocomplement of a is denoted by a' and it satisfies

$$a \subset b \Rightarrow b' \subset a'$$
$$a \cap a' = \phi$$
$$a \cup a' = I,$$

where ϕ is the smallest and I the largest element in the lattice. These elements take the role of the null set and the entire set in the classical case.

Two propositions $a, b \in \mathscr{L}$ are said to be *disjoint* if $a \subset b'$ where b' is the orthocomplement of b. In this form the definition of disjointness is identical with the classical one. The notation for this relation is also $a \perp b$. The lattice has a smallest and a largest element denoted by ϕ and by I, respectively. In a sense the role of Ω in the classical case is taken now by the element $I \in \mathscr{L}$.

A probability measure on $\mathscr{L}$ is a function $\mu : \mathscr{L} \rightarrow [0, 1]$ defined on $\mathscr{L}$ with values in $[0, 1]$ satisfying the following conditions

(i) $\Sigma \mu(a_i) = \mu(\bigcup a_i)$ for $a_i \in \mathscr{L}$, $i = 1, 2, \ldots, a_i \perp a_k$ for $i \neq k$.

(ii) $\mu(\phi) = 0$, $\mu(I) = 1$.

(iii)   If $\mu(a) = \mu(b) = 1$ then $\mu(a \cap b) = 1$.

The first two properties are exactly as in the classical calculus; the third is new. In fact in the classical calculus the third is a consequence of the other two. In the quantal calculus it is independent and therefore has to be postulated separately.

Passing now to the definition of random variables we use the definition which does not refer to the space Ω.

DEFINITION. *A random variable is a σ-homomorphism $\xi : \mathfrak{B}(\mathbb{R}) \rightarrow \mathscr{L}$ from the Borel sets on the real line into the lattice $\mathscr{L}$ of propositions, which satisfies the following conditions.*

(i) $\xi(\phi) = \phi$, $\xi(\mathbb{R}) = I$.

(ii)   *For any disjoint sequence $\Delta_i \in \mathfrak{B}(\mathbb{R})$ ($\Delta_i \perp \Delta_k$, for $i \neq k$)*
        $\xi(\bigcup \Delta_i) = \bigcup_i \xi(\Delta_i)$.

(iii) $\Delta_1 \perp \Delta_2 \Rightarrow \xi(\Delta_1) \perp \xi(\Delta_2)$.

An immediate consequence of these properties is that the range of the map ξ is a Boolean sublattice of $\mathscr{L}$. This is due to the fact the map is a homomorphism, that is, it conserves the lattice structure, which means that

$$\xi(\Delta_1 \cup \Delta_2) = \xi(\Delta_1) \cup \xi(\Delta_2)$$
$$\xi(\Delta_1 \cap \Delta_2) = \xi(\Delta_1) \cap \xi(\Delta_2)$$
$$\xi(\Delta') = \xi(\Delta)'.$$

From this follows that the image of the map is a Boolean sublattice of $\mathscr{L}$. The distribution function $F_\xi(a)$ is defined, as before, by

$$F_\xi(a) = \mu(\xi(S_a)) \quad \text{with} \quad S_a = (-\infty, a].$$

It induces a Stieltjes-Lebesgue measure μ_ξ on the Borel sets Δ.

The expectation value of a random variable ξ is then defined by

$$\langle \xi \rangle = \int\limits_{-\infty}^{+\infty} a \, dF_\xi(a).$$

From the foregoing we see that there is a close analogy between the classical and the quantal probability calculus. But there is also a profound difference due to the fact that the lattice of yes-no experiments for a quantal system is non-Boolean. The difference becomes explicit when we study the notion of *joint probability distribution* of two random variables.

It is useful to begin with the notion of compatibility. Two elements $a, b \in \mathscr{L}$ are said to be *compatible* and we denote this relation with $a \leftrightarrow b$ if the smallest sublattice which contains a, b, a', and b' is Boolean. We call this the lattice *generated* by a and b.

It is easy to see that a sublattice $\mathfrak{B} \subset \mathscr{L}$ is Boolean if and only if every pair of elements from $\mathfrak{B}$ is compatible.

The notion of compatibility can be transferred to random variables. To this end we define the ranges

$$\mathfrak{B}_\xi = \{a \mid a = \xi(\Delta), \Delta \in \mathfrak{B}(\mathbb{R})\}$$
$$\mathfrak{B}_\eta = \{a \mid a = \eta(\Delta), \Delta \in \mathfrak{B}(\mathbb{R})\}$$

and call ξ and η compatible if every $a \in \mathfrak{B}_\xi$ is compatible with every $b \in \mathfrak{B}_\eta$.

For pairs of classical random variables one can define the notion of *joint distribution*. It is defined as follows: let ξ and η be two classical random variables. The joint distribution is a function of two real variables a and b,

$$F_{\xi,\eta}(a, b) = \mu(\xi(S_a) \cap \eta(S_b)).$$

It is a nondecreasing function of both arguments satisfying the further conditions:

(i) $\qquad F_{\xi,\eta}(-\infty, b) = F_{\xi,\eta}(a, -\infty) = 0.$

(ii) $\qquad \displaystyle\int_{-\infty}^{+\infty} d_b F_{\xi,\eta}(a, b) = F_\xi(a) = \mu(\xi(S_a)).$

(iii) $\qquad \displaystyle\int_{-\infty}^{+\infty} d_a F_{\xi,\eta}(a, b) = F_\eta(b) = \mu(\eta(S_b)).$

Since compatible random variables in the quantum probability calculus behave exactly like classical ones it is immediately obvious that such variables also have a joint probability distribution given by formulas identical with the preceding ones.

VI. RANDOM VARIABLES IN HILBERT SPACE

Before discussing the question of the distribution function of noncompatible random variables in quantum probability calculus we give the interpretation of random variables in Hilbert space.

It is known that every proposition system $\mathscr{L}$ admits a representation in a linear vector space with coefficients from the real, complex, or quaternion fields. This representation is particularly simple if the lattice is irreducible, or in physical terms, if the system admits no superselection rules.

The mathematical expression for this property is that the center $\mathscr{C}$ of $\mathscr{L}$ is trivial. With the *center* $\mathscr{C}$ we denote the set of elements which are compatible with every other element:

$$\mathscr{C} = \{a \mid a \in \mathscr{L}, a \leftrightarrow x, \forall x \in \mathscr{L}\}.$$

Evidently $\phi \in \mathscr{C}$ and $I \in \mathscr{C}$. If these are the only two elements contained in $\mathscr{C}$ then we refer to $\mathscr{C}$ as being trivial.

The subspaces (or the projection operators) in a Hilbert space $\mathscr{H}$ form a lattice with $\phi = 0$ ($=$ zero projection), $I = I$ ($=$ unit operator) $E' = I - E$ (orthocomplement), and $I \cap F = S - \lim_{n \to \infty} (EF)^n$ ($=$ meet). The join is then defined by $E \cup F = (E' \cap F')'$. Under some mild additional restrictions one can show that the coefficients of the Hilbert space are the complex number field C.[11] We shall assume that this is the case. Under these hypotheses the abstract lattice of propositions is isomorphic to the lattice of subspaces of a Hilbert space $\mathcal{H}$, as it was demonstrated by Piron (1964).

Let us now examine what becomes of a probability measure and random variables in this case.

Let E_i $(i = 1, 2, ...)$ be a sequence of pairwise disjoint projections ($E_i \perp E_k$ or equivalently $E_i E_k = 0$ for $i \neq k$), then a probability measure is a functional μ from the set of all projections $\mathcal{P}$ to the interval $[0, 1]$

$$\mu : \mathcal{P} \to [0, 1]$$

satisfying the three characteristic properties

(i) $\bigcup_i \mu(E_i) = \mu(\sum E_i)$.

(ii) $\mu(\phi) = 0, \qquad \mu(I) = 1$.

(iii) $\mu(E) = \mu(F) = 1 \Rightarrow \mu(E \cap F) = 1$.

According to a theorem due to Gleason,[12] if $\dim \mathcal{H} \geqslant 3$ every such measure can be represented by a positive trace class operator ρ of trace 1, such that

$$\mu(E) = \operatorname{Tr} \rho E.$$

In the special case that ρ is a projection operator of rank 1 we have $\rho^2 = \rho$ and if φ is in the range of ρ, so that $\rho \varphi = \varphi$, one obtains

$$\mu(E) = (\varphi, E\varphi).$$

In this manner we recover the usual expectation values for pure states as they occur in quantum mechanics.

Let us now consider a random variable in this setting. According to the definition of Part V, a real random variable is a σ-homomorphism $\xi : \mathfrak{B}(\mathbb{R}) \to \mathcal{P}$ from the Borel sets on the real line to the projections in $\mathcal{H}$, which satisfies the three conditions (i), (ii), and (iii) given in Part V.

An inspection of these conditions shows that these are exactly the conditions for the definition of a *spectral measure*. According to the spectral theorem every spectral measure defines uniquely a self-adjoint operator X according to the formula

$$X = \int_{-\infty}^{+\infty} \lambda \, dE_\lambda$$

with

$$E_\lambda = \xi((-\infty, \lambda]) \equiv \xi(S_\lambda).$$

From Gleason's theorem follows then that the expectation value of ξ in the state μ is given by

$$\langle \xi \rangle = \mathrm{Tr}\,\rho X = \int_{-\infty}^{+\infty} \lambda \, d\,\mathrm{Tr}(\rho E_\lambda).$$

Thus we have recovered all the usual formulas of quantum theory in Hilbert space.

I add a few comments to this result.

(1) I stated that property (iii) of the probability measure must be postulated since it cannot be derived from the other two as in the classical probability calculus. In order to appreciate this remark, I sketch the derivation of (iii) from the other two conditions in the classical case when $\mathscr{L}$ is a Boolean algebra.

THEOREM 1. *If $\mathscr{L}$ is a Boolean algebra, and μ is a function μ: satisfying conditions* (i) *and* (ii) *then*

$$\mu(a) = \mu(b) = 1 \Rightarrow \mu(a \cap b) = 1 \; \forall a, b \in \mathscr{L}.$$

Proof. If $a \cap b = \phi$ then they are disjoint. Hence by (i) $\mu(a) + \mu(b) = \mu(a \cup b) = 1$. Therefore $\mu(a) = 1 \Rightarrow \mu(b) = 0$ and $\mu(b) = 1 \Rightarrow \mu(a) = 0$. The hypotheses of the theorem cannot be satisfied.

We may thus assume that $a \cap b = c \neq \phi$. We may then write

$$a = a_1 \cup c$$
$$b = b_1 \cup c$$

where $a_1 = c' \cap a$, $b_1 = c' \cap b$, and a_1, b_1, c are pairwise disjoint. Hence from (i) we obtain

$$1 = \mu(a) = \mu(a_1 \cup c) = \mu(a_1) + \mu(c)$$
$$1 = \mu(b) = \mu(b_1 \cup c) = \mu(b_1) + \mu(c).$$

By taking the difference of these two questions we find first that

$$\mu(a_1) = \mu(b_1) \equiv x.$$

On the other hand from the sum of the two questions we obtain

(1) $\qquad 1 = x + \mu(c)$

since $1 = \mu(a) \leqslant \mu(a \cup b) \leqslant 1$ we have $\mu(a \cup b) = 1$ and therefore from (i)

$$1 = \mu(a \cup b) = \mu(a_1) + \mu(b_1) + \mu(c)$$

or

(2) $\qquad 1 = 2x + \mu(c).$

Comparing (1) with (2) we conclude that $x = 0$ and therefore

$$\mu(a \cap b) = 1 \quad \|.$$

(2) In the Hilbert space setting property (iii) can actually also be proved as a consequence of (i) and (ii) provided $\dim \mathscr{H} \geqslant 3$.

This is due to the following facts:

(a) Under this hypothesis every probability measure μ is of the form $\mu(E) = \mathrm{Tr}\,\rho E$ with ρ a positive trace class operator with trace 1;

(b) If E, F are any two projections then

$$E \cap F = S - \lim_{n \to \infty} (EF)^n;$$

(c) If T_n is a uniformly bounded sequence of operators and $T_n \to T$ strongly, then $\mathrm{Tr}\,\rho T$ exists and

$$\mathrm{Tr}\,\rho T_n \to \mathrm{Tr}\,\rho T,$$

where (a) is essentially Gleason's theorem quoted in this part, (b) is a well-known result on projections in Hilbert space (cf. note 1), (c) can be proved as follows: the operator ρ being of trace class may be written as

$$\rho = \sum_{r=1}^{\infty} \alpha_r P_r$$

where P_r are orthogonal projections which we may assume without loss of generality to be of rank 1.

The eigenvalues α_r may be ordered as a decreasing sequence

$$\alpha_1 \geqslant \alpha_2 \geqslant \cdots \geqslant \alpha_r \geqslant \cdots \geqslant 0.$$

Furthermore the trace condition means $\sum \alpha_r = 1$. Let $R < \infty$ be an integer such that $\sum_{R+1}^{\infty} \alpha_r < \varepsilon$ for some arbitrary $\varepsilon > 0$, and let $P_r \varphi_r = \varphi_r$, $\|\varphi_r\| = 1$. We obtain then for

$$|\mu(T_n) - \mu(T)| = |\sum_R \alpha_r (\varphi_r, (T_n - T) \varphi_r)|$$
$$\leqslant \sum_{r=1}^{R} \alpha_r |(\varphi_r, (T_n - T) \varphi_r)| + \sum_{R+1}^{\infty} \alpha_r |(\varphi_r, (T_n - T) \varphi_r)|.$$

We now choose N such that for $n > N$

$$|(\varphi_r, (T_n - T) \varphi_r)| < \varepsilon \quad \forall (r = 1, 2, ..., R).$$

This is possible because $T_n \to T$ strongly, hence weakly. The first term becomes therefore

$$< \varepsilon \sum_{r=1}^{R} \alpha_r < \varepsilon.$$

For the second term we note that because $T_n \to T$ and T_n are uniformly bounded, T is also bounded, hence $|(\varphi_r, (T_n - T) \varphi_r)| \leqslant \|T_n\| + \|T\|$, so that the second term is

$$\leqslant (\sum_{R+1}^{\infty} \alpha_r)(\|T_n\| + \|T\|) \leqslant \varepsilon(\|T_n\| + \|T\|).$$

Because of the uniform boundedness the right-hand side is independent of n. Hence we have shown

$$\mathrm{Tr}\, \rho T \quad \text{exists an} \quad \mathrm{Tr}\, \rho T = \lim_{n \to \infty} \mathrm{Tr}\, \rho T_n.$$

Let us now verify property (iii). We note first that

$$\mathrm{Tr}\, \rho E = \sum_{r=1}^{\infty} \alpha_r (\varphi_r, E\varphi_r) = 1$$

implies

$$(\varphi_r, E\varphi_r) = \|E\varphi_r\|^2 = 1 \quad (r = 1, 2, ..., \infty),$$

so that

$$\|\varphi\|^2 = 1 = \|E\varphi\|^2 + \|(I - E)\varphi\|^2$$

or

$$\|(I - E)\varphi\|^2 = 0, \quad \text{or finally} \quad E\varphi = \varphi.$$

Thus for all r such that $\alpha_r > 0$

$$E\varphi_r = \varphi_r.$$

Similarly

$$F\varphi_r = \varphi_r.$$

Therefore

$$\sum \alpha_r(\varphi_r, EF\varphi_r) = \sum \alpha_r = 1,$$

so that

$$\mathrm{Tr}\,\rho EF = 1.$$

we denote $EF = T$, we note that $\|T^n\| \leqslant 1$ and conclude from the preceding reasoning that

$$\mathrm{Tr}\,\rho T^n = 1 \quad (n = 1, 2, \ldots).$$

Hence by (c)

$$\mu(E \cap F) = \mathrm{Tr}\,\rho E \cap F = 1 \quad \|.$$

(3) Property (iii) has a simple physical interpretation in case there exist passive filters corresponding to the propositions a and b. Indeed $\mu(a) = 1$ says that the filter corresponding to a is 100 percent transparent. Similarly $\mu(b) = 1$ implies that the filter corresponding to b is also 100 percent transparent. Since the filters are passive the system traverses the filters without modification of the state. Hence it will also traverse an infinite (or very large) alternating sequence of filters a and b. But such a sequence represents the filter corresponding to $a \cap b$. Hence $\mu(a \cap b) = 1$.

The only example known to me of a probability measure on a lattice which does not satisfy (iii) is in a lattice with a maximal chain of three elements. This is of course precisely the case that is excluded by the hypothesis of Gleason's theorem that $\dim \mathcal{H} \geqslant 3$. In view of this fact it would be of considerable interest to prove property (iii) in the lattice-theoretic setting. No such proof is known to me.

(4) The present derivation of Hilbert-space quantum theory from the lattice-theoretic one elucidates the relation between *compatibility* of observables and *commutativity* of the corresponding operators in Hilbert space.

The former is a physical property and the latter a mathematical one. In the light of the phenomenological interpretation of the lattice structure, compatibility is represented by the relation $a \leftrightarrow b$ which in turn is equivalent to the property that the sublattice generated by (a, b, a', b') is Boolean. This is exactly how it is in classical physics, where every such sublattice is Boolean since the entire proposition system $\mathscr{L}$ is.

In the representation of the proposition system by the subspaces of a Hilbert space, compatibility of two projection operations E, F is equivalent to commutability of these operators. We have in fact the following.

THEOREM 2. $\mathscr{L}(E, F, E', F')$ is Boolean $\Leftrightarrow [E, F] = 0$.

Proof. If $\mathscr{L}(E, F, E', F')$ is Boolean then

$$E = E_1 + G$$
$$F = F_1 + G$$

with

$$G = E \cap F, \qquad E_1 = E \cap G', \qquad F_1 = F \cap G'.$$

It follows that

$$EF = (E_1 + G)(F_1 + G) = G$$
$$FE = (F_1 + G)(E_1 + G) = G.$$

Therefore

$$[E, F] = 0 \quad \|.$$

If $[E, F] = 0$, then $E \cap F = EF$. Hence for any triplet, for instance E, F, F', we have

$$E \cap (F \cup E') = (E \cap F) \cup (E \cap E') = (E \cap F) \cup \phi = EF.$$

But

$$F \cup E' = I - E + FE,$$

so that

$$E \cap (F \cup E') = E(I - E + FE) = EFE = FE = EF.$$

Thus for any triplet chosen from E, F, E', F' we have the distributive law and this implies that $\mathscr{L}(E, F, E', F')$ is Boolean. $\quad \|$

This result disagrees with the opinion expressed by Park and Margenau in a recent publication (see Park and Margenau, 1968). However it is

seen that this result is independent of the hypothesis whether the correspondence between observables and self-adjoint operators is one to one
contrary to what is claimed in that reference. In fact the essential hypothesis is the much weaker one that with the propositions a, $b \in \mathscr{L}$ the
proposition 'a and b' $= a \cap b$ is also contained in $\mathscr{L}$.

VII. JOINT DISTRIBUTIONS OF RANDOM VARIABLES

In the new quantum probability calculus there is an essential feature
which distinguishes it from the classical calculus. This is the occurrence
of noncompatible observables or random variables. In the classical
calculus every observable is compatible with every other one, due to the
fact that the lattice $\mathscr{L}$ is Boolean. In the quantum calculus this is not
necessarily the case.

In the classical case it was possible to define the joint distribution
function of two random variables ξ, η as the function $F_{\xi,\eta}(a, b)$ satisfying
the following properties

$$(0) \qquad F_{\xi,\eta}(a, b) \geqslant 0 \quad \text{and nondecreasing in } a \text{ and } b.$$

$$(1) \qquad F_{\xi,\eta}(-\infty, b) = F_{\xi,\eta}(a, -\infty) = 0.$$

$$(2) \qquad \int_{-\infty}^{+\infty} d_b F_{\xi,\eta}(a, b) = F_\xi(a).$$

$$(3) \qquad \int_{-\infty}^{+\infty} d_a F_{\xi,\eta}(a, b) = F_\eta(b).$$

In the quantal case the definition of such a joint probability may be
impossible in case the random variables ξ and η are not compatible.
This corresponds to the physical fact that joint measurements of arbitrary
noncompatible variables may be impossible.

Since the preceding statement is flatly contradicted by Park and
Margenau,[13] I must interpose at this point a few critical remarks concerning their analysis of the measuring process in quantum theory.

Their analysis concerns primarily the notion of pairs of incompatible
observables. They insist that in spite of the uncertainty relation, such as
$\Delta p \Delta q \geqslant \frac{1}{2}\hbar$ for canonical variables p and q, such variables can be measured

with arbitrary degree of accuracy. They therefore reject complementarity, so essential in Bohr's analysis of the quantal systems, and they believe this concept can be replaced by the simpler notion of 'latency'. Although they would agree that the uncertainty relation is valid for measurements on an ensemble of identically prepared systems, they believe that this relation is not a restriction concerning the accuracy of measurements for complementary variables of an individual system.

The essential point in their analysis concerns the 'joint' measurements of noncommuting observables. Although an explicit definition of their notion of compatibility is never given in their paper one gathers from the context that for them compatibility means that a joint measurement of the pair of observables is possible. By showing that certain pairs of noncommuting observables are measurable simultaneously to an arbitrary degree of accuracy they come to the conclusion that noncommuting observables may very well be compatible in their sense of the term. (Incidentally it is not clear from their paper whether they believe that any pair of noncommuting observables is compatible in this sense or not.)

They conclude from this that there are joint measurements possible for certain variables such as p and q even though neither a joint probability distribution nor an operator exists for representing such joint measurements.

In order to appraise this point of view it is necessary to recall that in Bohr's point of view the arbitrary precision of individual measurements of canonical variables such as p and q was never in question. Both quantities can in principle be measured with a precision only limited by the inherent precision of the applied experimental arrangement. However the very presence of this experimental arrangement precludes the simultaneous attributions of precisions to complementary variables, such as p and q which would violate the uncertainty relation.

The example given by Park and Margenau for such a measurement is no counterexample to this general and essential feature of quantal systems. Their example is in fact only a determination of the position q with a given accuracy Δq followed by a determination of p a long time t later with an arbitrary accuracy Δp. Their conclusion that this second measurement permits them to assert that this also constitutes a measurement of p with that same accuracy at time $t=0$ is not correct. Their justification for this is that the probability distribution of p at time $t=0$ is the same as at a time $t>0$. While this statement is perfectly correct it is not sufficient for asserting that the actual value of the p at the two times is equal.

With this counterexample shown to be irrelevant for the question under discussion their entire case falls to the ground and the difficulties which they had to face concerning joint distributions of noncommuting observables disappear.

Returning now to the problem of the joint probability distribution for incompatible observables it is very easy to see in the quantum probability calculus that such a distribution cannot exist satisfying the properties listed above for the canonical variables.

The reason for this is the fundamental relation

(1) $\qquad \xi(S_a) \cap \eta(S_b) = \phi \quad$ for $\quad -\infty < a, b < +\infty$.

Indeed if $\xi(S_a) \cap \eta(S_b)$ were $\neq \phi$ then there would exist a function $\varphi(x) \in L^2(-\infty, +\infty)$ with the properties

$$\varphi(x) = 0 \quad \text{for} \quad x > a$$
$$\hat{\varphi}(x) = 0 \quad \text{for} \quad x > b$$

where $\hat{\varphi}$ is the Fourier transform of φ. It is well known that such a function does not exist unless $\|\varphi\| = 0$.

Due to the relation (1) it follows that

$$F_{\xi, \eta}(a, b) = 0 \quad \text{for} \quad -\infty < a, b < +\infty$$

so that properties (2) and (3) are violated.

But this negative conclusion does not preclude that p and q (or in fact any pair of noncommuting observables) are measurable within an accuracy limited by the uncertainty relation. Thus a joint probability distribution should exist in a more restrictive sense which is in accord with this restriction.

In order to define this weaker sense we modify the definition of $F_{\xi, \eta}$. Instead of a nondecreasing function on $\mathbb{R}^2$ we define it as a finitely additive set function on the Borel rectangles. If ξ and η are compatible then this definition is possible and the function $F_{\xi, \eta}$ satisfies the properties

(0) $\qquad F_{\xi, \eta}(A \times B) \geqslant 0$

$$(1) \quad F_{\xi,\eta}(\phi \times B) = F_{\xi,\eta}(A \times \phi) = 0$$
$$(2) \quad F_{\xi,\eta}(A \times \mathbb{R}) = F_{\xi}(A) = \mu(\xi(A))$$
$$(3) \quad F_{\xi,\eta}(\mathbb{R} \times B) = F_{\eta}(B) = \mu(\eta(B)).$$

According to a well-known theorem in measure theory such a finitely additive set function on Borel rectangles has a unique extension to the Borel sets on $\mathbb{R}^2$, defining a product measure on $\mathbb{R}^2$.

For noncompatible random variables, such as p and q, it is still possible to define a function $F_{\xi,\eta}$ satisfying all the properties listed above by setting

$$F_{\xi,\eta}(A \times B) = \mu(\xi(A) \cap \eta(B)).$$

But in agreement with Wigner's (1932) result this function is not an additive set function on Borel rectangles and therefore cannot be extended to a measure on $\mathbb{R}^2$.

In spite of this anomaly the function $F_{\xi,\eta}(A \times b)$ is not entirely devoid of physical meaning. It represents in fact the probability that in a given state the variable ξ assumes values in the set A while at the same time the variable η assumes values in the set B.

This probability is not necessarily zero as may be seen in the case $\xi = p$ and $\eta = q$. In this case as we have noted before $\xi(A) \cap \eta(B) = \phi$ if $m(A') m(B') = \infty$.

But in case $m(A') m(B') < \infty$ this is not true. We have in fact the following.

THEOREM 3. *If $E \equiv E_A$, $F \equiv F_B$ represent the spectral projections of the canonical variables p and q associated respectively with the Borel sets A and B, then*

$$m(A') m(B') < \infty \Rightarrow E \cap F \neq \phi.$$

The proof of this theorem will be given elsewhere. Suffice it here to remark that the theorem implies the following statement concerning functions $\varphi \in L^2(-\infty, +\infty)$ and their Fourier transform $\hat{\varphi}$. We shall say that φ has a gap (of positive measure) if there exists a Borel set A with Lebesgue measure $m(A)$ such that

$$0 < m(A') < \infty.$$

The theorem then asserts that there exist functions $\varphi(x)$ with a gap whose Fourier transform also has a gap.[14]

In conclusion I may thus state that, although joint measurements of certain noncompatible observable may be possible with nontrivial results, it is not true that there exists an observable which represents all the joint measurements of two such observables.

University of Geneva

NOTES

* This article was written during a visit to the University of Colorado in Boulder. It is a pleasure to thank Professor K. Gustafson, who made this visit possible. Thanks are also due to Professor B. Misra for an important remark concerning the joint distribution of noncompatible observables. This study originally appeared in *Synthese* 29 (1974), 131–154.

[1] The quantum probability calculus was briefly sketched by Jauch (1968). The role of probability in quantum theory was the principal subject of three papers by Suppes (1961, 1963, 1966).

[2] For a detailed review of hidden variable theories the reader is referred to Belinfante (1973).

[3] This point was first made by Wigner (1932). It was the subject of many subsequent papers such as Moyal (1949), Brittin and Chappell (1962), Park and Margenau (1968).

[4] For recent papers on this subject I refer to Gudder (1967, 1968), Gudder and Marchand (1972), and Varadarajan (1962).

[5] The principal difference with respect to some other work in this field is that the quantal proposition system is assumed to be a lattice and not just an orthocomplemented partially ordered set (poset). The empirical justifications for this assumption were first given by Piron (1964), where it was shown quite explicitly that for many physical systems the poset structure is not sufficient for representing the phenomenology. Further details are given by Jauch (1968).

[6] This useful distinction is due to my late friend, Dr. G. Baron, whose profound knowledge of fundamental problems on probability theory has greatly influenced my thinking on the subject.

[7] Kolmogorov (1956). This is the English version of the original German version.

[8] Renyi (1970) introduced the probability calculus based on the basic notion of relative probability. This generalizes Kolmogorov's calculus to nonnormalizable probability fields.

[9] This form of quantum probability calculus was first developed by Varadarajan (1962).

[10] Jauch (1968) uses for instance the construction of composite filters. There are other possibilities of constructing the meet of two elementary noncompatible events.

[11] The question of the number field remained for a long time beyond an empirical test. The recent work by Gudder and Piron (1971) is the best that one can do.

[12] The conjecture that every σ-additive measure on orthogonal subspaces is given by a density matrix was finally proved by Gleason (1957). Similar conjectures on the projection lattice of von Neumann algebras remain unproved.

[13] Park and Margenau (1968) claim to have shown that measurements are possible which violate the uncertainty relations.

[14] I am indebted to Prof. Martin Peter for an explicit construction of such functions,

who also showed that their existence is not without physical interest especially in the theory of metals. The question whether such functions exist and their relevance for the problem of joint distributions was first mentioned to me by Prof. A. Galindo of Madrid.

BIBLIOGRAPHY

Belinfante, F. J., *A Survey of Hidden Variable Theories*, Pergamon Press, New York, 1973.
Brittin, W. E. and Chappell, W. R., 'The Wigner Distribution Function and Second Quantization in Phase Space', *Review of Modern Physics* **34** (1962), 620–627.
Gleason, A. M., 'Measures on the Closed Subspaces of a Hilbert Space', *Journal of Mathematics and Mechanics* **6** (1957), 885–893.
Gudder, S. P., 'Hilbert Space, Independence, and Generalized Probability', *Journal of Mathematical Analysis and Applications* **20** (1967), 48–61.
Gudder, S. P., 'Joint Distributions of Observables', *Journal of Mathematics and Mechanics* **18** (1968), 325–335.
Gudder, S. P. and Marchand, J. P., 'Non-Commutative Probability on von Neumann Algebras', *Journal of Mathematical Physics* **13** (1972), 799–806.
Gudder, S. P. and Piron, C., 'Observables and the Field in Quantum Mechanics', *Journal of Mathematical Physics* **12** (1971), 1583–1588.
Jauch, J. M., *Foundations of Quantum Mechanics*, Addison-Wesley, Reading, 1968.
Kolmogorov, A. N., *Foundations of the Theory of Probability*, Chelsea, New York, 1956, transl. and ed. by N. Morrison.
Moyal, J. E., 'Quantum Mechanics as a Statistical Theory', *Proceedings of the Cambridge Philosophical Society* **45** (1949), 99–124.
Park, J. L. and Margenau, H., 'Simultaneous Measurability in Quantum Theory', *International Journal of Theoretical Physics* **1** (1968), 211–282.
Piron, C., 'Axiomatique Quantique', *Helvetica Physica Acta* **37** (1964), 439–467.
Renyi, A., *Probability Theory*, North Holland, Amsterdam, 1970.
Suppes, P., 'Probability Concepts in Quantum Mechanics', *Philosophy of Science* **28** (1961), 378–389.
Suppes, P., 'The Role of Probability in Quantum Mechanics' in B. Baumrin (ed.), *Philosophy of Science, The Delaware Seminar*, Wiley, New York, 1963.
Suppes, P., 'The Probabilistic Argument for a Non-classical Logic of Quantum Mechanics', *Philosophy of Science* **33** (1966), 14–21.
Varadarajan, V. S., 'Probability in Physics and a Theorem on Simultaneous Observability', *Comments on Pure and Applied Mathematics* **15** (1962), 189–216.
Wigner, E., 'On the Quantum Theory for Thermodynamic Equilibrium', *Physical Review* **40** (1932), 749–759.

G. CASSINELLI AND E. G. BELTRAMETTI

QUANTUM LOGICS AND IDEAL
MEASUREMENTS OF THE FIRST KIND

We choose the framework of a probabilistic interpretation of the states of a physical system: a state α is a probability measure on the set $\mathscr{L}$ of propositions (classes of equivalent yes-no experiments) whose minimal structure is that of orthoposet containing the union of disjoint elements. We denote by $\mathscr{S}$ the set of the states. Let $a \in \mathscr{L}$, $\alpha \in \mathscr{S}$; then $\alpha: \mathscr{L} \to [0, 1]$, and $\alpha(a)$ is physically interpreted as the probability of the yes response of a when the initial state of the system is α.

The probabilistic interpretation of the states has been adopted by many authors: as typical examples we might quote Mackey and Pool[1, 2]. A non-probabilistic definition of pure states has been advanced by Jauch and Piron in [3]; we shall not enter into the discussion of this point.

The probability function $\alpha \in \mathscr{S}$ provides a 'passive' description of the physical system, since it considers only the occurrence of the yes response of $a \in \mathscr{L}$. To get an 'active' description of the physical system one should consider also the study of the state of the system after the measurement of a. That opens the way to introducing conditional probabilities, and of making the link with the prescriptions of the quantum theory of measurement. The problem arises whether the lattice structure of $\mathscr{L}$ has relations with the transformation properties of the state of the system under the measurement of $a \in \mathscr{L}$. In this note we shall examine to what extent the lattice structure of $\mathscr{L}$ determines these transformation properties. What we do is, roughly speaking, dual of the approach suggested in [2] by Pool some years ago: let us, first, give a summary of it.

By a set of physically interpretable axioms, Pool adopts a basic proposition-state structure $(\mathscr{L}, \mathscr{S})$, where $\mathscr{L}$ is an orthomodular poset containing the union of disjoint elements and $\mathscr{S}$ a strongly ordering, σ-convex set of probability measures on $\mathscr{L}$. The ordering relation and the orthocomplementation in $\mathscr{L}$ are, explicitly,

J. Leite Lopes and M. Paty (eds.), Quantum Mechanics, a Half Century Later, 63–67. All Rights Reserved.

$$a \leqslant b \Leftrightarrow \{\alpha \in \mathcal{S}: \alpha(a) = 1\} \subseteq \{\alpha \in \mathcal{S}: \alpha(b) = 1\} \quad a, b \in \mathcal{L},$$

$$a \mapsto a^{\perp} \text{ where } \{\alpha \in \mathcal{S}: \alpha(a^{\perp}) = 1\} = \{\alpha \in \mathcal{S}: \alpha(a) = 0\} \quad a, a^{\perp} \in \mathcal{L}.$$

Then it is assumed that each $a \in \mathcal{L}$ determines uniquely a mapping Ω_a of $\mathcal{S}$ into $\mathcal{S}$ which is physically interpreted as the transformation of the state of the system caused by the measurement of a. More precisely, if the initial state of the system is α and if the yes response of a occurs, then the system is left in the final state $\Omega_a(\alpha)$. The mapping Ω_a is assumed to have the properties of an ideal measurement of the first kind and some further properties which make the set

$$S_\Omega = \{\Omega_{a_1} \circ \Omega_{a_2} \circ \cdots \circ \Omega_{a_n}: a_1, a_2, \ldots, a_n \in \mathcal{L}\},$$

a Baer-*semigroup (semigroup with respect to the operation $\circ$ of composition of maps). With this structure, one can make use of the remarkable connections between Baer-*semigroups and orthomodular lattices (for definition and properties of Baer*-semigroups we refer to Maeda and Maeda and Pool [4, 2]). $\mathcal{L}$ is recognized as the set of the closed projections of S_Ω, so that one can deduce, in particular, that

(i) $\mathcal{L}$ is an orthomodular lattice, and the ordering is equivalent to $\Omega_a \circ \Omega_b = \Omega_a$;

(ii) the following commutativity relations between a and b are equivalent

(1) $\Omega_a \circ \Omega_b = \Omega_b \circ \Omega_a$,

(2) $a = (a \wedge b) \vee (a \wedge b^{\perp})$,

(3) $(a, b, a^{\perp}, b^{\perp})$ generate a Boolean sub-lattice.

Moreover, Pool shows that the covering law (see, e.g. [4]) of $\mathcal{L}$ can be based on some hypothesis about supports of states and their transformation properties under the mapping Ω_a, $a \in \mathcal{L}$.

Let us recall that $a \in \mathcal{L}$ is said to be the support of $\alpha \in \mathcal{S}$, and we write $a = \sigma(\alpha)$, when

$$\alpha(b) = 0 \Leftrightarrow a \perp b.$$

We now come to our main aim: to reverse that approach and see whether the orthomodular lattice structure of $\mathcal{L}$ contains in itself information about the active picture of the propositions.

Our starting mathematical structure is a pair $(\mathcal{L}, \mathcal{S})$ which fulfils the following

AXIOM. $\mathscr{L}$ is a complete orthomodular lattice, $\mathscr{S}$ is a strongly ordering, σ-convex set of probability measures on $\mathscr{L}$. If $\alpha \in \mathscr{S}$ then the support of α exists in $\mathscr{L}$. If $a \in \mathscr{L}\backslash\{\underline{0},\underline{1}\}$, then there exists $\alpha \in \mathscr{S}$ such that $a = \sigma(\alpha)$.

Hence a surjective mapping is defined

$$\sigma: \mathscr{S} \to \mathscr{L}\backslash\{\underline{0},\underline{1}\}.$$

It is known that an orthomodular lattice is always isomorphic to the lattice of the closed projections of some Baer-*semigroup[2, 4]. The set $S(\mathscr{L})$ of the residual mappings (or emimorphisms) of $\mathscr{L}$ is proved to be a Baer-*semigroup[2, 4]; $\mathscr{L}$ is isomorphic to the lattice $P'(S(\mathscr{L}))$ of the closed projections of $S(\mathscr{L})$: precisely, for every $a \in \mathscr{L}$,

$$\varphi_a(b) = (b \vee a^\perp) \wedge a, \quad b \in \mathscr{L},$$

is an element of $P'(S(\mathscr{L}))$; conversely, every element of $P'(S(\mathscr{L}))$ is of the form φ_a for some $a \in \mathscr{L}$.

Given any $\alpha \in \mathscr{S}$, consider its support $\sigma(\alpha)$ and transform it by φ_a, $a \in \mathscr{L}$. If $\sigma(\alpha) \mathcal{L} a$, or, equivalently, if $\alpha(a) \neq 0$, the proposition

$$\varphi_a(\sigma(\alpha)) = (\sigma(\alpha) \vee a^\perp) \wedge a$$

belongs to $\mathscr{L}\backslash\{\underline{0},\underline{1}\}$. Hence the previous axiom ensures that there exists at least one state β whose support is $\varphi_a(\sigma(\alpha))$. This suggests the following result, proved in [5]: for every $a \in \mathscr{L}$, there exists at least one mapping $\Omega_a: \mathscr{S} \to \mathscr{S}$ which makes commutative the diagram,

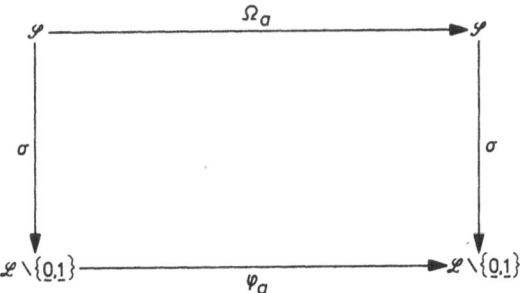

and has domain

$$\mathscr{D}[\Omega_a] = \{\alpha \in \mathscr{S}: \alpha(a) \neq 0\}.$$

Moreover, such mapping Ω_a satisfies the following properties ([5]):

(A) $\alpha \in \mathcal{D}[\Omega_a], \quad \alpha(a) = 1 \Rightarrow \Omega_a(\alpha) = \alpha,$

(B) $\alpha \in \mathcal{D}[\Omega_a], \quad \beta = \Omega_a(\alpha) \Rightarrow \beta(a) = 1,$

(C) $\alpha \in \mathcal{D}[\Omega_a], \quad \beta = \Omega_a(\alpha), b \in \mathcal{L}, a \, C \, b,$

(D) $\alpha \in \mathcal{D}[\Omega_a], \quad \beta = \Omega_a(\alpha), b \in \mathcal{L}, a \, C \, b \Rightarrow \beta(b) = \beta(a \wedge b),$

(E) $\alpha \in \mathcal{D}[\Omega_a], \quad \beta = \Omega_a(\alpha) \Rightarrow \sigma(\beta) = \varphi_a(\sigma(\alpha)),$

where the commutativity relation C is here understood in the sense of orthomodular lattice, i.e.,

$$a \, C \, b \quad \text{whenever} \quad a = (a \wedge b) \vee (a \wedge b^{\perp}).$$

We adopt for Ω_a the following physical interpretation: if the initial state of the system is α, the interaction with the apparatus used to measure a leaves the system in the final state $\Omega_a(\alpha)$ whenever the yes response occurred. Accordingly, the form of $\mathcal{D}[\Omega_a]$ shows that Ω_a is not defined on the states which give with certainty the no response of a. The properties (A) and (B) correspond to the definition of a measurement of the first kind: the state of the system is left unchanged if the yes response of $a \in \mathcal{L}$ is certain, and the repetition of the experiment a will give with certainty the yes response. The property (C) corresponds to the commonly accepted definition of ideal measurement: if a and b commute, the measurement of b does not disturb the measurement of a. These results show that the $(\mathcal{L}, \mathcal{S})$ structure (equipped with the previous axiom) is sufficient to deduce the existence of ideal measurements of the first kind. Moreover, when $a \, C \, b$ and β is a state filtered by a, the property (D) suggests for $\beta(b)$ the interpretation of conditional probability of compatible events (to guarantee formally this interpretation one should further require:

$$a \geqslant b, \quad \beta = \Omega_a(\alpha) \Rightarrow \beta(b) = \frac{\alpha(b)}{\alpha(a)}).$$

The property (E), which determines uniquely the support of $\Omega_a(\alpha)$ from the support of α, is the essential step to equip $\mathcal{L}$ with the covering law.

These properties of the mapping Ω_a, deduced from the $(\mathcal{L}, \mathcal{S})$ structure, coincide (up to some technical points) with assumptions occurring in Pool's approach. We miss just one significant point: Ω_a is not uniquely determined by a. This lack of uniqueness lies in the fact that the mapping $\sigma\colon \mathscr{S} \to \mathscr{L}\backslash\{\underline{0}, \underline{1}\}$ (see the previous diagram) is surjective, not bijective.

However, the problem of the uniqueness of Ω_a is connected with further hypotheses, about atomicity of $\mathscr{L}$ and pure states of $\mathscr{S}$, which are needed to introduce the covering law. For the last we refer to the definition: $\mathscr{L}$ has the covering law if $\varphi_a(p)$ is an atom for every $a \in \mathscr{L}$ and for every atom $p \in \mathscr{L}$ such that $p \not\leq a^{\perp}$.

In fact, assume, as usual, the hypothesis: $\mathscr{L}$ is atomic, pure states exist in $\mathscr{S}$, and σ determines a bijection between the atoms of $\mathscr{L}$ and the pure states of $\mathscr{S}$. Then by use of the property (E), it follows that among the statements

    I. $\mathscr{L}$ has the covering law,

    II. the restriction of Ω_a to the pure states of $\mathscr{D}[\Omega_a]$ is uniquely determined by a,

    III. Ω_a is a pure operation, i.e., transforms pure states into pure states,

the following implications hold:

$$I \Rightarrow II \text{ and } III,$$

$$III \Rightarrow I \text{ and } II.$$

We conclude that also the connections between covering law and pure operations are contained, in a natural way, into the $(\mathscr{L}, \mathscr{S})$ structure.

Istituto di Scienze Fisiche dell' Università
di Genova (Italy)

BIBLIOGRAPHY

[1] Mackey, G. W., *The Mathematical Foundations of Quantum Mechanics*, Benjamin, New York, 1963.
[2] Pool, J. C. T., *Commun. Math. Phys.* **9** (1968), 118.
    Pool, J. C. T., *Commun. Math. Phys.* **9** (1968), 212.
[3] Jauch, J. M. and Piron, C., *Helv. Phys. Acta* **42** (1969), 842.
    Jauch, J. M., 'Foundations of Quantum Mechanics', in *Proc. Int. School of Phys.* 'Enrico Fermi', 49° corso. Academic Press, New York-London, 1971.
[4] Maeda, F. and Maeda, S., *Theory of Symmetric Lattices*, Springer Verlag, Berlin-Heidelberg-New York, 1970.
[5] Cassinelli, G. and Beltrametti, E. G., *Commun. Math. Phys.* **40** (1975), 7.

C. PIRON

A FIRST LECTURE ON QUANTUM MECHANICS

ABSTRACT. In this paper we give, on an introductory level, a unified formulation of 'quantum' physics. The formalism which is obtained by taking seriously Einstein's point of view and describe a physical system in terms of 'elements of reality', is presented in the spirit of what might be called the 'school of Geneva'. The lecture is divided into four sections. First we introduce the notions of physical system, question and proposition, and show that the propositions are naturally embedded in a lattice. In the second section we discuss the propositional system, introducing the postulates of 'quantum' physics. Furthermore, we define the notion of state of a physical system. Next, we define what is meant by saying that two propositions are compatible and introduce some criteria for compatibility. We also consider some realizations of propositional systems, before ending the section by defining the concept of observable and justify the use of self-adjoint operators. The last section is devoted to a proof of the probability law of quantum mechanics, usually assumed as a postulate.

I. DESCRIPTION OF PHYSICAL SYSTEMS

The goal of a physical theory is to describe and predict the results of possible experiments which are performed on physical systems. By a physical system we understand a part of reality conceived as existing in space-time and exterior to the physicist. The precise definition of a physical system always depends on the point of view and the degree of idealization considered.

Each affirmation of a physicist relative to a physical system should admit experimental testing. This experiment consists in a measurement, the result of which is expressed by 'yes' or 'no'. If the answer is 'yes', the affirmation of the physicist is confirmed but not proved. If the answer is 'no' the affirmation of the physicist is refuted and is therefore *false*.

To affirm that a physical system possesses a certain property, means that a particular test which one could possibly carry out would give this property with certainty.

DEFINITION. A *question* is an experiment leading to an alternative of which the terms are 'yes' or 'no'.

Thus a question is both a description of an experiment to be carried

J. Leite Lopes and M. Paty (eds.), Quantum Mechanics, a Half Century Later, 69–87. All Rights Reserved

out on the physical system considered and a rule enabling us to interpret the possible results in terms of 'yes' or 'no'.

More schematically, a question consists of
  (i) a measuring apparatus,
 (ii) instruction for its use,
(iii) a rule interpreting the possible results in terms of 'yes' or 'no'.

Let us illustrate the previous definition by an example.

Example E1

Physical system: a chalk.

Question: This chalk is breakable.
  (i) The measuring apparatus is the two hands.
 (ii) The experiment consists of taking the chalk with the two hands and bending it with the maximum force.
(iii) If the chalk is broken, the answer is 'yes', otherwise 'no'.

One can define a great number of questions for a given physical system. Particularly with each question α one can associate an inverse question denoted $\alpha^\sim$.

DEFINITION.   Given a question α, we can define *the inverse question*, the question $\alpha^\sim$ obtained by exchanging the terms of the alternative.

For example, the inverse question of E1 is obtained by modifying (iii) as follows:
(iii) The answer is 'no' if the chalk is broken and it is 'yes' in all other cases.

There exists a *trivial question*, denoted as I, which consists in doing anything (or nothing) with the system considered and stating the answer 'yes' each time. The question $I^\sim$ is the absurd question, denoted 0.

Given a family of questions α_i, we can define a product question $\Pi_i \alpha_i$, in the following manner:

DEFINITION. $\Pi_i \alpha_i$ is the question defined as follows:
  (i) the apparatus is the set of apparatuses of α_i.
 (ii) one performs an arbitrary one of the α_i, and not necessarily always the same,
(iii) the result obtained is attributed to $\Pi \alpha_i$.

If the family $\{\alpha_i\}$ is composed of only two elements α and β, one writes $\alpha \cdot \beta$ for the product.

DEFINITION.   If the physical system has been prepared in such a way that in the case where one performs a question α, the answer 'yes' will be *certain*, we shall say that the question α is 'true' for the physical system.

Remark.   If the question α is not true, it does not happen, in general, that $\alpha^{\sim}$ is 'true'.

DEFINITION.   We say that a question α is stronger than a question β, i.e. $\alpha < \beta$, if:

α 'true' $\Rightarrow \beta$ 'true'.

This means that the answer 'yes' for β is certain when the answer 'yes' for α is certain.

This relation is *transitive* and permits us to define an equivalence relation denoted by $\sim$.

$\alpha \sim \beta$ if $\alpha < \beta$ and $\beta < \alpha$.

It leads to the following definition:

DEFINITION.   A '*proposition*' is an equivalence class of questions.

When a question α is 'true' for a given physical system, all the questions of the class of equivalence containing α are also 'true' for the same system and we say that '*the proposition a is true*' (for this particular system).

Let $\mathscr{L}$ be the set of propositions relative to a given physical system.

The set $\mathscr{L}$ is provided with an *order relation*. We write $a < b$, a and $b \in \mathscr{L}$, if the questions of a are stronger than those of b, i.e. if a 'true' implies b 'true'.

Let us show that for every family of propositions of $\mathscr{L}$ one can define a least upper bound and a greatest lower bound.

Let $\{a_i\}$ be a non-empty family of elements of $\mathscr{L}$. We denote by $\bigwedge_i a_i$ the equivalence class containing the question $\Pi_i \alpha_i$ where $\alpha_i \in a_i \, \forall_i$. This proposition is independent of the choice of α_i in the equivalence class a_i. Thus:

   (i) $\bigwedge_i a_i < a_i, \quad \forall i$,

(ii) $\forall\, x \in \mathscr{L}$ such that, $x < a_i$, $\forall\, i$ we have $x < \wedge_i\, a_i$.

(i) is an immediate consequence of the fact that $\Pi_i\, \alpha_i$ is 'true' if each α_i is true and it shows that the proposition $\wedge_i\, a_i$ is a lower bound for $\{a_i\}$.

(ii) results from the following fact. Let ξ be a question of x. Then ξ 'true' $\Rightarrow \alpha_i$ 'true', $\forall\, i \Rightarrow \Pi_i\, \alpha_i$ 'true'. This shows that $x < \wedge_i\, a_i$.

(ii) implies that $\wedge_i\, a_i$ is the greatest lower bound of the a_i.

The least upper bound of the family of proposition $\{a_i\}$ can now be defined as the greatest lower bound of the upper bounds of the a_i. Let $\mathscr{F}\{x \in \mathscr{L}\,|\,a_i < x,\ \forall\, i\}$ be the set of upper bounds of the a_i. Then

$$\bigvee_i a_i = \bigwedge_{x \in \mathscr{F}} x$$

defines the greatest upper bound of a_i. Let us remark that the set is never empty because it always contains the trivial Proposition I.

DEFINITION. We call a *complete lattice* a set provided with an order relation for which each sub-set admits a greatest lower bound and a least upper bound.

From the previous considerations it follows that the set of propositions relative to a given physical system is a *complete lattice*.

It is apparent from the previous discussion that a 'true' and b 'true' $\Leftrightarrow a \wedge b$ 'true'.

So that lower bound $\wedge$ plays an analogous role to the *and* of logic. On the other hand, as regards the upper bound, we have:

a 'true' or b 'true' $\Rightarrow a \vee b$ 'true'.

THEOREM. If $a \vee b$ 'true' $\Rightarrow a$ 'true' or b 'true', then $\mathscr{L}$ is a distributive lattice.

Proof. (i) $a \wedge (b \vee c)$ 'true' $\Leftrightarrow a$ 'true' and $b \vee c$ 'true' $\Leftrightarrow a$ 'true' *and* (b 'true' *or* c 'true') $\Leftrightarrow a \wedge b$ 'true' or $a \wedge c$ 'true' $\Leftrightarrow (a \wedge b) \vee (a \wedge c)$ 'true'.

Thus when $\vee$ plays an analogous role to the 'or' of logic, the lattice $\mathscr{L}$ is distributive. We shall then say that it is classical because in fact the lattice of the propositions of a classical physical system in the usual sense is always distributive.

II. SYSTEMS OF PROPOSITIONS.
STATES OF A PHYSICAL SYSTEM

We have ascertained that the set $\mathscr{L}$ of the propositions relative to a given physical system constitute a complete lattice, and we will continue in this section to exhibit some of its particular properties.

DEFINITION. The complement of an element $a \in \mathscr{L}$ is defined as an element $b \in \mathscr{L}$ such that:

$$a \wedge b = 0 \quad \text{and} \quad a \vee b = I.$$

This definition is evidently symmetric.

When the lattice $\mathscr{L}$ is distributive, the complement, if it exists, is unique. In fact, let $a \in \mathscr{L}$ be an element possessing two complements b_1 and b_2. Then $b_2 = b_2 \wedge (a \vee b_1) = (b_2 \wedge a) \vee (b_2 \wedge b_1) = b_2 \wedge b_1 = b_1$, by symmetry.

DEFINITION. We call a compatible complement of an element $a \in \mathscr{L}$ a complement b of a for which there exists a question $\alpha \in a$ such that $\alpha^{\sim} \in b$.

Remark. When b is a compatible complement of a, the existence of a question $\alpha \in a$ like $\alpha^{\sim} \in b$ permits one to raise the alternative:

$$a \text{ 'true' or } b \text{ 'true'}$$

in a single experiment.

The existence of a compatible complement is a general property of the proposition lattice of the physical system, which gives rise to the following postulate:

POSTULATE I

For every proposition $a \in \mathscr{L}$, there exists at least one *compatible complement.*

We denote by a' a compatible complement of $a \in \mathscr{L}$.

The lattice $\mathscr{L}$ of propositions of a non-classical system does not differ radically in its structure, from a distributive lattice in the sense that it contains distributive sub-lattices.

POSTULATE II

    For all elements a, $b \in \mathscr{L}$ such that $a < b$, the sub-lattice generated by the elements a, b, a' and b' is *distributive*.

The following properties result from Postulate II:

(i) $a < b \Rightarrow b \wedge (a \vee b') = a$, $\forall a$ and $b \in \mathscr{L}$,

(ii) $a = (a')'$, $\forall a \in \mathscr{L}$,

(iii) $a < b \Rightarrow b' < a'$, $\forall a$ and $b \in \mathscr{L}$.

DEFINITION. A lattice $\mathscr{L}$ is said to be a *CROC* if it is provided with a mapping which with each element $a \in \mathscr{L}$ associates an element $a' \in \mathscr{L}$, called the *orthocomplement*, in such a way that:

(i) $(a')' = a$,

(ii) $a \wedge a' = 0$ and $a \vee a' = I$,

(iii) $a < b \Rightarrow b' \vee (a' \wedge b) = a'$.

THEOREM. Let $\mathscr{L}$ be a CROC. If we consider the orthocomplements as compatible complements, then $\mathscr{L}$ satisfies Postulates I and II.

We then proceed by defining the concept of state of a physical system.

In the case of a classical physical system, in the usual sense, the state defines all the actual properties of the system.

We want the state to characterize all the actual properties of a physical system, and this justifies the following definition:

DEFINITION. We will define the state of a given physical system as the set $E \subset \mathscr{L}$ of all the propositions actually 'true' for this system.

This definition is consistent with the interpretation of the structure of the lattice $\mathscr{L}$ only if E satisfies the following conditions:

(i) $0 \notin E$ and $I \in E$,

(ii) $a \in E$, $a < x \Rightarrow x \in E$,

(iii) if $\{a_i\}$ is a family of elements of E, then we have $\wedge_i a_i \in E$.

It results from the conditions (i) and (iii) that a state E possesses a greatest lower bound $\neq 0$ which is: $p = \Lambda_{x \in E} x$.

On the other hand the greatest lower bound $p \in E$ characterizes the state E, since by virtue of (ii) we have:

$$E = \{x \in \mathscr{L} | p < x\}.$$

Let us suppose that there exists an element $z \neq 0$ and p such that $z < P$. The existence of such an element would then have, as a consequence, the possibility of modifying the state E of the system considered, in such a way that the proposition z would be also 'true' for this system.

That is the reason for which we impose on the state E the following additional condition:

(iv) $z < p$ and $z \notin E \Rightarrow z = 0$.

The condition (iv) explains the maximality of the physical state.

DEFINITION. We call an atom, an element $p \neq 0$ of $\mathscr{L}$ such that $z < p \Rightarrow z = 0$ or $z = p$.

Thus the set of all possible states of a physical system is in one-to-one correspondence with the set of atoms of the corresponding lattice of propositions. This gives rise to the following properties, which we write down as two postulates:

POSTULATE III

$\forall\, a \in \mathscr{L}$ such that $a \neq 0$, $\exists$ an atom $p < a$. We shall say that the lattice $\mathscr{L}$ is *atomic*.

In fact if $a \neq 0$, there exists a state of the system for which a is true, therefore there exists an atom $p < a$.

POSTULATE IV

$\forall\, a \in \mathscr{L}$ and $\forall$ atom $p \in \mathscr{L}$ such that $a' \wedge p = 0$, $(p \vee a') \wedge a$ is an *atom*.

The significance of this fourth postulate will be clearly explained in Section IV.

DEFINITION. A complete lattice of propositions which satisfies Postulates I, II, III, and IV is called a *system of propositions.*

Example E6

The set $\mathscr{P}(E)$ of all the subsets of a set E is an example of a distributive system of propositions.

In fact, $\mathscr{P}(E)$ is a complete lattice and it is a CROC for the mapping which associates with a subset of E its set-complement. The atoms of $\mathscr{P}(E)$ are the points of E. Finally, it is trivial to verify that Postulates III and IV are satisfied.

In this way, in the case of a classical physical system described by a phase space Γ, the corresponding system of propositions is defined by the lattice $\mathscr{P}(\Gamma)$ of all subsets of Γ. The corresponding *states* are then the *points* of Γ.

Example E7

The set $\mathscr{P}(H)$ of closed sub-space of a *Hilbert* space H is a non-distributive system of propositions.

In fact, the set $\mathscr{P}(H)$ is a complete lattice, which is a *CROC* for the mapping which associates to a sub-space F of H its orthogonal complement $F^{\perp}$, since for F and $G \in \mathscr{P}(H)$.

(i) $(F^{\perp})^{\perp} = F$,

(ii) $F \cap F^{\perp} = \{0\}$ and $F + F^{\perp} = H$ (Rietz theorem),

(iii) $G \subset F \Rightarrow F^{\perp} + (G^{\perp} \cap F) = G^{\perp}$ (Generalized Rietz theorem).

The atoms of $\mathscr{P}(H)$ are the one-dimensional sub-spaces of H, i.e. the rays.

Let P be an atom of $\mathscr{P}(H)$ such that $P \cap F^{\perp} = \{0\}$, then $(P + F^{\perp}) \cap F$ is also an atom (even in the case of infinite dimensional H, since $F^{\perp}$ is closed, $P + F^{\perp}$ is also closed).

III. COMPATIBILITY AND OBSERVABLES

As we have previously mentioned, the system of propositions $\mathscr{L}$ of a physical system possesses distributive sub-lattices. More particularly according to Postulate II, each couple of elements a and b of $\mathscr{L}$ such that $a < b$ generates with their respective compatible complements a' and b', a distributive sub-lattice. The first part of this section is devoted to the study of some of these sub-lattices.

DEFINITION.   We call a sub-CROC, a sub-set $\mathscr{B}$ of $\mathscr{L}$ such that:

(i) $a_i \in \mathscr{B}, \ \forall \, i \in J \Rightarrow \bigvee_{i \in J} a_i \in \mathscr{B}$,

(ii) $a \in \mathscr{B} \Rightarrow a' \in \mathscr{B}$.

Let $\{\mathscr{B}_i\}$ be a family of a sub-CROC of $\mathscr{L}$. Their intersection (in the sense of the set-theory) $\bigcap_i \mathscr{B}_i$ is again a sub-CROC.

Given a sub-set $\mathscr{A}$ of $\mathscr{L}$, there exists a smallest sub-CROC containing $\mathscr{A}$ which is the sub-CROC generated by $\mathscr{A}$, i.e. the intersection of all the sub-CROCs of $\mathscr{L}$ which contain $\mathscr{A}$.

DEFINITION.   We shall say that a proposition $a \in \mathscr{L}$ is compatible with the proposition $b \in \mathscr{L}$ (we shall note $a \leftrightarrow b$) if the sub-CROC generated by a and b is distributive.

According to Postulate II, if $a < b$, then $a \leftrightarrow b$. If $a < b'$ we have $a \leftrightarrow b$ equally.

In Example E7, if F and $G \in \mathscr{P}(H)$ and if $F \subset G^\perp$, then $F \leftrightarrow G$. Geometrically F and G are orthogonal.

$$(F \perp G \text{ if: } x \in F \quad \text{and} \quad y \in G \Rightarrow x \perp y).$$

Thus we shall say that $a \in \mathscr{L}$ is orthogonal to $b \in \mathscr{L}$, and we denote it by $a \perp b$, when $a < b'$.

In Example E6, it is easy to verify that each proposition of $\mathscr{P}(E)$ is compatible with each proposition of $\mathscr{P}(E)$.

There exist many algebraic characterizations of the compatibility of two propositions, specially the following one, which we give without proof.

Criteria of compatibility

$$a \leftrightarrow b \Leftrightarrow a \wedge (b \vee a') = a \wedge b.$$

The relation $a \leftrightarrow b$ is no doubt symmetric, but it is not *transitive*; if $a \leftrightarrow b$ and $b \leftrightarrow c$, a is not necessarily compatible with c.

On the other hand, we have the following two theorems:

THEOREM.   Let $a \leftrightarrow b_i, \ \forall \, i \in J$. Then:

$$a \leftrightarrow \bigwedge_{i \in J} b_i \quad \text{and} \quad a \leftrightarrow \bigvee_{i \in J} b_i.$$

THEOREM.   Let $\{a_i\}$ be a family of propositions of $\mathscr{L}$ which are *compatible* two by two. Then the sub-CROC generated by that family of propositions is *distributive*.

According to the last theorem, a physicist content in doing only the experiment concerning a family of propositions, compatible two by two, would see that the physical system considered exhibited all properties of a classical system.

We intend now to display the particular form taken by compatibility in the case of the system of propositions of Example E7. This study makes use of the *orthogonal projector* of the Hilbert space H.

The projectors have the following properties:

(i) $P_F = P_F^+,$

(ii) $F \subset G \Leftrightarrow P_F = P_G P_F \Leftrightarrow P_F = P_F P_G,$

(iii) $F \perp G \Leftrightarrow P_F P_G = 0 \Leftrightarrow P_G P_F = 0 \Leftrightarrow P_{F+G} = P_F + P_G.$

THEOREM.   Let F and G be two elements of $\mathscr{P}(H)$ and let P_F and P_G be corresponding projectors. Then:

$$F \leftrightarrow G \Leftrightarrow [P_F, P_G] = 0.$$

Proof. If $F \Leftrightarrow G$, we have $F = (F \cap G) + (F \cap G^\perp)$ because the sub-CROC generated by F and G is distributive. Since $F \cap G \perp F \cap G^\perp$ we can write: $P_F = P_{F \cap G} + P_{F \cap G^\perp}$, by virtue of the Property (iii) of the projectors. But $F \cap G \subset G$ and $G \perp F \cap G^\perp$ imply $[P_G, P_{F \cap G}] = 0$ and $[P_G, P_{F \cap G^\perp}] = 0$ according to (ii) and (iii). Consequently, $[P_F, P_G] = 0$.

Reciprocally, if $[P_F, P_G] = 0$, we can write:

$$P_{F \cap (F^\perp + G)} = P_F P_{F^\perp + G} = P_F (I - P_F + P_F P_G) = P_F P_G = P_{F \cap G},$$

from whence we have, $F \cap (F^\perp + G) = F \cap G$, which means that: $F \Leftrightarrow G$, according to the criteria of compatibility.

Accordingly the compatibility of the two elements F and $G \in \mathscr{P}(H)$ for the system of propositions of Example E7, is equivalent to the commutativity of the corresponding projectors P_F and P_G.

Commutativity is characterized geometrically by the following fact:

When P_F and P_G commute, the Hilbert space H decomposes into the sum of the orthogonal sub-spaces $F \cap G$, $F \cap G^\perp$, $F^\perp \cap G$ and $F^\perp \cap G^\perp$. The corresponding projectors are $P_F P_G$, $P_F - P_F P_G$, $P_G -$

$P_F P_G$ and $I - P_F - P_G + P_F P_G$ respectively, which can be verified without difficulty.

DEFINITION. We call the *'centre'* of a system of propositions the set $\mathscr{Z}$ of the propositions of $\mathscr{L}$ which are compatible with all of the propositions of $\mathscr{L}$.

$$\mathscr{Z} = \{z \in \mathscr{L} \mid z \leftrightarrow a, \ \forall \, a \in \mathscr{L}\}.$$

According to the first two theorems of this section, the centre $\mathscr{Z}$ of a system of propositions $\mathscr{L}$ is a distributive sub-CROC. One can even show that it is a sub-system of propositions.

We are now in position to explain the difference between classical physical system and quantum physical system.

A system of propositions $\mathscr{L}$ is said to be *purely classic* if its centre $\mathscr{Z}$ is identical with $\mathscr{L}$ and *purely quantal* if its centre is $\{0, I\}$.

There exists intermediate physical systems; for historical reasons, one says that such a system possesses *superselection rules*.

A purely classical system of proposition can be realized by all the subsets of a set (Example E6); and in general, a purely quantal system of propositions can be realized by the closed sub-spaces of a Hilbert space (Example E7).

In the general case, the system of propositions can be realized in the following way:

Realization in the general case
We consider a family $\{H_\omega\}$ of the Hilbert space H_ω indexed by the elements ω of a set Ω. The set of propositions is then realized by the families $\{P_\omega\}$ of projectors of H_ω. The order relation is defined by:

$$\{P_\omega\} < \{Q_\omega\} \Leftrightarrow P_\omega Q_\omega = P_\omega, \ \forall \, \omega \in \Omega,$$

and the compatible complement is given by

$$\{P_\omega\}' = \{I - P_\omega\}.$$

The lattice thus defined satisfies Postulates I, II, III, and IV. In particular, it is easy to verify that the atoms of $\mathscr{L}$ are the families of projectors, all null except one which is of rank 1.

A state is thus characterized by an element $\omega_0 \in \Omega$ and by a ray of H_{ω_0}.

The compatibility is given by:

$$\{P_\omega\} \leftrightarrow \{Q_\omega\} \Leftrightarrow [P_\omega, Q_\omega] = Q_\omega, \ \forall_\omega \in \Omega.$$

Finally the *centre* $\mathscr{Z}$ of this system of propositions is the set of families $\{P_\omega\}$ for which $P_\omega = 0$ or 1, $\forall \omega \in \Omega$.

Each element of $\mathscr{Z}$ is defined by the sub-set of $\omega \in \Omega$ for which $P_\omega = 1$; thus the centre $\mathscr{Z}$ of $\mathscr{L}$ is isomorphic to $P(\Omega)$. We say then that Ω is *superselection rules*.

The end of this section is devoted to the mathematical notion of observable.

The role of the measuring apparatus is essentially to indicate to the physicist whether such or such proposition concerning the physical system considered, is true.

As a result, a measuring apparatus implies a certain kind of correspondence between the distributive CROC $\mathscr{B}$ of the propositions concerning its scale and some propositions of $\mathscr{L}$.

Mathematically, this correspondence is realized by a mapping of the distributive CROC $\mathscr{B}$ into the system of propositions $\mathscr{L}$ which preserves the lattice structure and the orthogonality.

The previous arguments suggest the following definition:

DEFINITION. Let $\mathscr{L}$ be a system of propositions. We call an observable a mapping $\mu \colon \mathscr{B} \to \mathscr{L}$, where $\mathscr{B}$ is a distributive CROC, and such that
  (i) $\mu(V_i a_i) = V_i \mu(a_i)$,
  (ii) $a \perp b \Rightarrow \mu(a) \perp \mu(b)$.
In general we also impose $\mu(I) = I$.

THEOREM. Let $\mathscr{L}_1$ and $\mathscr{L}_2$ be two CROCS. A mapping $\mu \colon \mathscr{L}_1 \to \mathscr{L}_2$ which preserves the least upper bound and orthogonality possesses also the following properties:
  (i) $\mu(0) = 0$,
  (ii) $\mu(a') = \mu(I) \wedge (\mu(a))'$,
  (iii) $\mu(V_i a_i) = V_i \mu(a_i)$.
Thus the image $\mu(\mathscr{B})$ of $\mathscr{B}$ in $\mathscr{L}$ is a distributive sub-CROC.

Conversely, each distributive sub-CROC defines an observable.

To conclude, we note that there does not necessarily exist a measuring apparatus corresponding to each mathematical observable.

Let us illustrate the notion of an observable by the following examples.

In the case (purely classical) of Example E6, every function $f: E \to D$, where D is some set, defines some observable $\mu: \mathcal{P}(D) \to \mathcal{P}(E)$ which with each $a \in \mathcal{P}(D)$ associates the element:

$$\mu(a) = f^{-1}(a) = \{x \mid x \in E, f(x) \in a\} \in \mathcal{P}(E).$$

Reciprocally one can show that each observable $\mu: \mathcal{P}(D) \to \mathcal{P}(E)$ defines a function $f: E \to D$ such that:

$$f^{-1}(a) = \mu(a), \quad \forall a \in \mathcal{P}(D).$$

In the case (purely quantal) of Example E7, if H is of finite dimension, it is easy to see that each self-adjoint *operator* A defines an observable.

Let us denote the set of eigen-values by $A = \Sigma_{\sigma(A)} \lambda_\gamma F_\gamma$.

The mapping μ of $\mathcal{P}(\sigma(A))$ into $\mathcal{P}(H)$ which with each $a \subset \mathcal{P}(\sigma(A))$ associates the sub-space $\mu(a) = \Sigma_{\lambda_\gamma \in a} P_\gamma$ is an *observable*.

Conversely let us consider an observable $\mu: \mathcal{P}(\sigma) \to \mathcal{P}(H)$ where σ is a finite sub-set of R. Let $\lambda_\gamma, \gamma = 1, \ldots, m$ be the elements of σ. Let $F_\gamma = \mu(\lambda_\gamma)$ be the sub-space of H associated with the element $\lambda_\gamma \in \sigma$. The sub-spaces F_γ for $\gamma = 1, \ldots, m$ are orthogonal amongst themselves and their sum is H. It is therefore easy to verify that the operator

$$A = \sum_{\gamma=1}^{m} \lambda_\gamma P_{F_\gamma},$$

is self-adjoint.

DEFINITION. We call *eigen-state* of the observable $\mu: \mathcal{B} \to \mathcal{L}$, a state $p \in \mathcal{L}$ when there exists an atom $e \in \mathcal{B}$ for which $p < \mu(e)$. e is called the *eigen-value* of the observable μ for the state p.

We would like to characterize the eigen-states of an observable.

If p is an eigen-state of μ then $p \leftrightarrow \mu(x), \forall x \in \mathcal{B}$. In fact let e be the eigen-value of μ for this state. Since $\mathcal{B}$ is a distributive CROC, we have

$$e = (x \wedge e) \vee (x' \wedge e).$$

Which gives: $e < x$ or $e < x'$. If $e < x$, then $p < \mu(e) < \mu(x)$ and if $e < x'$ then $p < \mu(e) < \mu(x') < (\mu(x))'$. In these two cases we have $p \leftrightarrow \mu(x)$.

DEFINITION. A state $p \in \mathcal{L}$ is said to be compatible with the observable $\mu: \mathcal{B} \to \mathcal{L}$ if $p \to \mu(\mathcal{B})$.

As we have just seen, an eigen state of the observable μ is compatible with μ. The converse is also true.

THEOREM. Let $\mu: \mathcal{B} \to \mathcal{L}$ be an observable such that $\mu(I) = I$ and let p be an atom of $\mathcal{L}$. If p is compatible with μ, then p is an eigen-state of μ.

We proceed by studying the structure of observables from this general point of view.

Let us consider an observable $\mu: \mathcal{B} \to \mathcal{L}$. Let $a \in \mathcal{B}$ and let the segment be:

$$[0; a] \equiv \{x \,|\, x \in \mathcal{B}, \, 0 < x < a\}.$$

This is a sublattice of $\mathcal{B}$ which is endowed with a structure of CROC by letting $x' \wedge a$ be the orthocomplement of x. The canonical injection:

$$\pi: [0; a] \to \mathcal{B},$$

preserves the least upper bound and the orthogonality. It thus follows that the composite mapping $\mu_0 \pi$ is a new observable, denoted by μ_a, the restriction of μ to the segment $[0; a]$.

Let $\mathrm{Ker}\,\mu$ be the *kernel* of μ.

$$\mathrm{Ker}\,\mu \equiv \{x \,|\, x \in \mathcal{B}, \, \mu(x) = 0\}.$$

The least upper bound $n = \mathrm{V}_{x \in \mathrm{Ker}\,\mu}\, x$ of the elements of $\mathrm{Ker}\,\mu$ belong to $\mathrm{Ker}\,\mu$ because $\mu(n) = \mathrm{V}_{x \in \mathrm{Ker}\,\mu}\, \mu(x) = 0$.

Finally, the observable μ is entirely defined by its restriction μ_n, to the segment $[0; n']$ and μ_n, is injective.

DEFINITION. By abuse of language the segment $[0; n']$ is called the *spectrum* of the observable μ. That spectrum is said to be:

    (i) purely discrete if $[0; n']$ is atomic,

    (ii) purely continuous if $[0; n']$ contains no atom.

In the *purely classical* case each observable $\mu: \mathcal{B} \to \mathcal{P}(E)$ has a *purely discrete* spectrum.

In fact if $a \neq 0$ is an element of the spectrum of μ, then $\mu(a) \neq 0$ and there exists an atom $p < \mu(a)$. Now $p \leftrightarrow \mu(\mathcal{B})$ since $\mathcal{L} \equiv \mathcal{P}(E)$ is purely classical and it follows from the previous results that p is an eigen-state of μ and that there exists an atom $e \in \mathcal{B}$ such that $p < \mu(e)$. But $p < \mu(e) \wedge \mu(a) = \mu(a \wedge e) \neq 0$ which implies $e < a$.

In short, every observable $\mu \colon \mathcal{B} \to \mathcal{P}(E)$ is defined by the inverse image of a function of E in the atoms of $\mathcal{B}$.

Remark. For an observable having a purely continuous spectrum there does not exist an eigen-state.

In the purely quantal case where the Hilbert space H is of finite dimension, the spectrum of an observable $\mu \colon \mathcal{B} \to \mathcal{P}(H)$ is a lattice which is finite and therefore atomic, and the spectrum of μ is *purely discrete*.

In the purely quantal case where H is separable, each self-adjoint operator defines an observable and conversely.

In fact, let A be a self-adjoint operator of H. We can write A in the following form:

$$A = \int \lambda P(\mathrm{d}\lambda), \qquad \lambda \in R,$$

where the projectors $P(\mathrm{d}\lambda)$ define the *spectral family* of the operator A.

The corresponding observable to the operator A is defined by the sub-CROC of the projectors of the spectral family and conversely each observable defines a spectral family and thus a self adjoint operator.

DEFINITION. The observables $\mu_i \colon \mathcal{B}_i \to \mathcal{L}$ are said to be compatible if the propositions of the images $\mu_i(\mathcal{B}_i)$ are compatible.

Finally, in the general case of system of propositions realised by a family of separable Hilbert spaces H_ω, ($\omega \in \Omega$), each family $\{A_\omega\}$ of self-adjoint operators defines an observable and conversely. Two observables $\{A_\omega\}$ and $\{B_\omega\}$ are compatible if:

$$[A_\omega, B_\omega] = 0, \quad \forall \, \omega \in \Omega.$$

IV. THEORY OF MEASUREMENT

Questions, as they have been defined in Section I, are experiments which in general modify completely the physical system considered, and sometimes even destroy it. Nevertheless, the definition of question permits us to establish the structure of the lattice of propositions.

However, there also exist experiments which try to satisfy the following conditions:

(1) to give information about the state,

(2) and this by perturbing the system as little as possible.

An experiment of this kind must permit one to conclude whether a certain proposition is true or not immediately after the experiment.

It is easily verified that for such experiments, there exists states which are perturbed whatever is the experimental technique involved. This is the case when the initial state of the system is not compatible with the proposition defined by the experiment.

Let us note that, the imperfections of the measuring apparatus may add to this natural inherent perturbation.

The *idealization* of the real experiment which approaches the previous conditions (1) and (2) leads to the notion of measurement which we define as follows:

DEFINITION. We shall call measurement a question $\alpha \in a$ if the answer 'yes' implies:

(i) a is 'true' immediately after the measurement,

(ii) x is 'true' immediately after the measurement if x is 'true' before and $x \leftrightarrow a$.

This idealization is necessary since it is desirable to give a description of the measuring process without taking into account the mechanisms particular to each measuring apparatus.

The solution of the following two problems is the main aim of this section.

(i) To determine the perturbation undergone by the system during a measurement if the answer is 'yes'.

(ii) To determine the probability of obtaining the answer 'yes' if the initial state of the system is known.

The solution of the problem (i) is given by the following theorem:

THEOREM. Let $\alpha \in a$ be a measurement satisfying the conditions (i) and (ii) of the preceding definition. If the measurement gives the answer 'yes', then the state of the system immediately after this measurement is

$$(p \lor a') \land a,$$

where p is the initial state of the system.

Proof. Let us suppose that the measurement α has given the answer 'yes'. Let x be any proposition compatible with a such that $p < x$. Then x is 'true' immediately after the measurement. Moreover $p < (a \vee p) \mid (a' \vee p) < (a \vee x) \wedge (a' \vee x) = x$.

Since the proposition $(a \vee p) \wedge (a' \vee p)$ is compatible with a (according to the first theorem of Section III), it is also true after the measurement. It is therefore the greatest lower bound of the propositions x. The proposition a being equally true after the measurement, it follows that

$$a \wedge [(a \vee p) \wedge (a' \vee p)] = (a' \vee p) \wedge a$$

is also true after the measurement. Finally, by Postulate IV $(a' \vee p) \wedge a$ is an atom. Since this atom characterizes the final state, the theorem is proved.

It is important to remark that without Postulate IV we would not be able to determine completely the final state and even when taking into account the answer of the system, the measurement would lose some information.

The previous theorem possesses a converse. If for any initial state p, a question α give as final state $(p \vee a') \wedge a$ when the answer is 'yes', then this question verifies the conditions (i) and (ii) of the definition of a measurement.

An immediate consequence of this theorem is the following fact: if the proposition a is initially 'true', then the measurement α does not disturb the system. The measurement α is therefore evidently an ideal experiment.

In the purely quantal case (Example E7), the condition on the system after a measurement concerning the proposition $F \in \mathcal{P}(H)$ which has obtained the answer 'yes', is according to the previous theorem

$$(G + F^{\perp}) \cap F,$$

where G is the ray of H characterizing the initial state. In other words, the final state is the projection of G on F, that is: $P_F G$.

Every non-zero vector of H characterizes a ray. Thus if ψ is a vector of G then:

$$P_F \psi \text{ characterizes the final state.}$$

In the same way, for a system of propositions realized by a family of Hilbert spaces H_ω, $\omega \in \Omega$, the initial state is given by a non-zero vector $\psi_{\omega_0} \in H_{\omega_0}$ and the final state by the vector

$$P_{\omega_0} \psi_{\omega_0} \in H_{\omega_0}$$

if the measurement relative to the propositions $\{P_\omega\}$ has given the answer 'yes'.

If the propositions $\{P_\omega\}$ and $\{Q_\omega\}$ are compatible, then the perturbation produced by the successive measurements of these two propositions does not depend on the order in which they are effected. The answer is twice 'yes'.

In fact we have:

$$P_{\omega_0} Q_{\omega_0} \psi_{\omega_0} = Q_{\omega_0} P_{\omega_0} \psi_{\omega_0},$$

for the final state of this measurement when the answer is twice 'yes', and the perturbation is *identical* to that which would be produced by the direct measurement of the proposition:

$$\{P_\omega\} \wedge \{Q_\omega\} = \{P_\omega Q_\omega\},$$

when the answer is 'yes'.

We will next consider the problem of determining the probability of obtaining the answer 'yes' for a given measurement, when we know the initial state.

HYPOTHESIS. The probability of obtaining the answer 'yes' for a system whose state is given, is the same for all the questions $\alpha \in a$ for which α or $\alpha^\sim$ satisfies the conditions (i) and (ii) in the definition of a measurement. This probability will be noted as $W_p(a)$.

LEMMA. We have:
  (i) $0 \leqslant W_p(a) \leqslant 1$, $W_p(0) = 0$ and $W_p(I) = 1$,
  (ii) $W_p(a) + W_p(a') = 1$,
  (iii) $p < a \Leftrightarrow W_p(a) = 1 \Leftrightarrow W_p(a') = 0$,
  (iv) $a \leftrightarrow b \Rightarrow W_{(p \vee a') \wedge a}(b) \cdot W_p(a) = W_{(p \vee b') \wedge b}(a) \cdot W_p(b) = W_p(a \wedge b)$.
  Proof. The properties (i), (ii) and (iii) result from the previous definitions and hypothesis. With respect to (iv), we have just seen that one can measure $a \wedge b$ by measuring successively a and b. From the laws of probability calculus, the probability of obtaining the answer 'yes' two times is given by the product

$$W_{(p \vee a') \wedge a}(b) \cdot W_p(a),$$

which is equal to $W_p(a \wedge b)$ according to the previous hypothesis.

With this we can apply Gleason's theorem, which permits us to determine the probability $W_p(a)$. We state the result without proof:

THEOREM. Let a system of proposition $\mathcal{L}$ realized by a family of complex Hilbert spaces H_ω, $\omega \in \Omega$ (such that $\forall \omega \in \Omega$, dim $H_\omega \neq 2$). Then there exists one and only one function $W_p(a)$ satisfying the conditions (i) to (iv) of the previous lemma. That function is given by

$$W_p(a) = \frac{\langle \psi_{\omega_0}, P_{\omega_0} \psi_{\omega_0} \rangle}{\langle \psi_{\omega_0}, \psi_{\omega_0} \rangle},$$

where ω_0 and ψ_{ω_0} represent the state p of the system and $\{P_\omega\}$ the proposition a.

Remark. The exceptional case where dim $H_\omega = 2$ never appears in practice because it is always possible to assume that such a system of propositions (as for example, that of the spin of the atom of Ag) is obtained by the restriction of a bigger system. It follows that in such a case the probability $W_p(a)$ as given by the previous theorem still holds.

University of Geneva, Switzerland

S. P. SHUSHURIN

ESSAY ON THE DEVELOPMENT OF THE STATISTICAL THEORY OF THE CALCULUS OF PROBABILITY*

ABSTRACT. This paper deals with a number of considerations related to the statistical theory of the probability calculus and its development; they can serve as a basis for the clarification of a number of problems in wave mechanics.

1. Statistical and probabilistic concepts are very important for both modern theoretical and experimental physics. The kinetic theory of gases provided the first application of statistics to physics, and thus opened up the new and important domain of statistical physics, of which statistical mechanics forms a part.

Statistical fluid dynamics (for instance the statistical theory of turbulence) constitutes a large part of fluid dynamics. The statistical theory of errors is required for each physical experiment, etc. . . . .

The statistical interpretation of wave mechanics, though not the only possible interpretation, has become very important[1].

A sudden break in the initial contributions can be seen when one analyses the beginnings of the statistical interpretation of wave mechanics made by Max Born. From his very first publication of 1926, Born[2] expounded a true statistical conception in speaking of electrons scattered on atoms. He introduced the 'exploitation' function (*Ausbeutungfunktion*) which corresponded to the characteristics of the atom considered as a 'black box'. According to this publication the wave function described the beam of electrons scattered on the atoms, i.e. it characterized the 'response' of the atom exposed to a monoenergetic beam of electrons.

A few months later, at a conference in Scotland in 1927, Born put forward another point of view: he considered the wave function as a means of describing probabilistically the evolution of the atom[3]. Such terms as 'statistical interpretation' and 'probabilistic interpretation' are still used as synonyms. But one might ask whether this identification of statistics and probability is correct. Since probability calculus is a highly developed branch of mathematics and very much used in physics, it would be useful to analyse its empirical foundations and its usefulness for physics.

J. Leite Lopes and M. Paty (eds.), Quantum Mechanics, a Half Century Later, 89–105. All Rights Reserved
Copyright © 1977 by S. P. Shushurin

2. Probability calculus, as a mathematical discipline, first originated in France, thanks mainly to the work of Pascal. Many French, English, German, and Russian mathematicians contributed to its development. The modern form of probability calculus was developed by P. S. Laplace, J. Bertrand, H. Poincaré, E. Borel, A. N. Kolmogorov, W. Feller, A. Renyi and some other younger mathematicians. It has often been emphasized that probability calculus, when considered as a set of game rules for the calculation of expected values, has a very extensive range. For instance H. Freudenthal, in his book *Probability and Statistics* [4] always speaks of 'the Addition Rule' or the 'Multiplication Rule'.

In the following we shall develop Mises' statistical concept as expounded in his works: *Wahrscheinlichkeitsrechnung* [5] and *Grundlagen des Wahrscheinlichkeitsrechnung* [6].

The disadvantage of Mises' concept is its neglect of the essential points in the analysis of empirical data. Mises did not give any strict definition of statistical law, in itself very important for probability calculus approach [7].

3. EMPIRICAL FOUNDATIONS OF STATISTICS

3.1. In the analysis of sets of data there are two stages of the empirical knowledge. The first stage is the classification of the data. Taxonomy began to develop as an offshoot of biology. Now it has become an important discipline for all the sciences not yet penetrated by the notion of quantity (namely: biology, medicine, psychology, archaeology, history, i.e. broadly speaking, all the human and social sciences) [8]. As an example of the simplest methods we could take Wroclaw taxonomy and the dendrites method, which were elaborated by a group of Polish archaeologists and mathematicians, led by the distinguished anthropologist Professor Czekanowski [9, 10].

It is the objective of taxonomical set analysis of empirical data to find a way of objectively partitioning a set into several subsets. Any set can always be classified arbitrarily but this has no research or scientific interest. In regrouping the various subsets, elements that are mutually close are always chosen. However this classification is not always clear-cut. Therefore the kind of discovery mentioned above is very valuable for any science.

The stage of the taxonomic analysis is less important for physics than for the other sciences mentioned above, because it was achieved as early as the old era of the development of physics. But physicists sometimes deal implicitly with taxonomic problems just as Molière's Mr. Jourdain implicitly spoke in prose. Every physical measurement is in fact a classification of objects according to the degree of any property. The problem of statistics in physics (classical, Bose-Einstein, and Fermi-Dirac statistics) is to choose different classifications for sets of micro-particles.

Two reasons led us to take up the taxonomic problem. Firstly conceptualization in each science demands the formation of notions, which would be impossible without a classification of empirical objects (either things or events). Secondly there exists an important difference between facts and laws, even at the taxonomic level of knowledge. Let us further explain this second point.

Consider a set of empirical objects already classified in some way. The result is a set of subsets of the initial set. This constitutes a *taxonomic fact*. If this fact is repeated several times, i.e. if a number of initial sets of the same empirical nature can be classified with the same number of subsets and if the characteristics of these sets are the same, we can infer the existence of a *taxonomic law*.

3.2. But the size (the number of elements) of the subsets can change when we go from the initial set to the others. To calculate the size of the subsets means the beginning of the statistical stage of the knowledge of the sets.

A set is supposed to be subdivided into several subsets. If the mutual intersection of each pair of these subsets is empty these subsets are called cells[11].

Now we introduce some basic definitions.

The number n_k of the elements of a cell number k is an *absolute part* of the cell number k. The relation $y_k = n_k/n$, in which n is the number of the elements of the initial set, is a *relative part* of the cell number k. A functional dependence of y_k in relation to k, or in relation to any parameter that characterizes the subset number k, can be constructed. It can be characterized by several parameters if the classification is multi-dimensional. This dependence is called the distribution density of an initial set over the set of the subsets (D.D.). D.D. can be represented graphically as a histogram or a polygon. The cumulative histogram of the distribution corresponds to the *distribution function* which is a common tool for mathematicians but not for physicists. If we obtain a D.D. as a result of the statistical analysis of an initial empirical set then we speak of *statistical fact*. If this fact on the whole is repeated, we can say that a *statistical law* has been discovered.[1]

One very important case of the empirical sets is represented by sets of events. They can be classified and statistical laws can be established for them. From these laws we may calculate more elaborate distributions for the sets obtained from these initial sets according to the rule of combinatorial algebra.

4. Valid results of knowledge are used as a basis for the rational action, in other words for working. But in order to obtain the desired result of a given action, we must make predictions. Thus the prediction (or the prognosis) is like a bridge between knowledge and the rational action.

The principles of scientific prediction are well established in the logical classical principle: "ab esse a posse valet, ab posse ad esse non valet" (the first part of the assertion is obviously the most important for the prediction).

Physics is a basic science since the predictions that flow from physical empirical laws are very accurate. Empirical physical laws have a functional form (functions of one or several scalar variables, or vector variables).

Predictions can also flow from taxonomical laws. This is usual in everyday life (you don't buy shoes at a dairy and vice versa). But statistical laws can be used too for prediction. The relative part of the subset of a set of events, that is, of a frequency of outcome, can be predicted by relying upon the empirical law established for sets of this kind. We call this predicted relative part as *probability* of the subset of events.

Probability calculus is thus defined as a scheme of quantitative predictions that rely on the empirical statistical laws (or upon the hypothetical statistical laws in order to verify or to falsify them). The notion of probability of an elementary event has no meaning. The probability of a subset of events can be spoken of only as the predicted relative part of this subset, as the probability of outcome.

This definition has a practical consequence: if statistical laws are accurate, then prediction is equally accurate, sound or valid.

5. In a few words we give below some important consequences for anybody willing to use the existing probability calculus.

5.1. The classical definition of probability is only a statistical definition at the taxonomical level. It has one defect: we can still add subsets that are empty from the empirical, statistical point of view, but that would change the value of Laplace's probability. Let us consider an example: the classical probability of outcome *heads* in the classic experiment of heads or tails with a coin is only of $\frac{1}{3}$, because there always exists the third theoretical outcome *side*. But it is always thought to be $\frac{1}{2}$ because the outcome side as subset of events is empty from the empirical statistical point of view, i.e. the third outcome never occurs in practice. If the width-to-diameter ratio of a coin is increased the corresponding subset ceases to be empty.

5.2. From the statistical point of view, the theorem of complete probabilities is somewhat ambiguous (the theorem on the multiplication of the independent exit probabilities is a peculiar case of the theorem of complete probabilities, quite used in physics). The answer, given by the theorem, is one of an infinity of possible true answers. (See Appendix I.)

5.3. Bernoulli's distribution in binomial form is a limit distribution similar to those of Poisson and Laplace-Gauss normal D.D. (See Appendix II.)

5.4. The simple generalization of the notion of probability in statistics allows one to pose, and to solve, many problems like Bertrand's paradoxes. The first and second Bertrand's paradoxes in generalized form were formulated and solved on the basis of the purely statistical theory of so-called 'geometrical probabilities', or, strictly speaking, of the statistical theory of continuous sets. (See Appendix III.)

5.5. The purely statistical theory of random processes can be easily developed. In particular, we obtained Einstein's law for the Brownian movement in a rigorous and elementary way. (See Appendix IV.) This approach has a pedagogic value since it can be more easily understood by undergraduate students and even by advanced pupils at secondary school level.

6. In conclusion we would like to express our hope that this approach might be helpful for the analysis of certain problems in modern wave mechanics.

APPENDIX I. ANALYSIS OF THE CONDITIONAL
PROBABILITIES THEOREM

Consider n balls in a box; m balls are black and $n - m$ are white. They can be classified according to their colours: One ball is taken out of the box, then put back into it; the balls are mixed carefully, then a second ball is taken out. Finally many such trials of two balls are performed. These trials can be classified according to various possible combinations of the colours. When we have two colours: black (B) and white (W), we have the Cartesian product of the set of colours, i.e. the set of elements BB, BW, WB, and WW.

Suppose the relative fractions corresponding to these four elements are respectively $\gamma_{BB} = x_1$, $\gamma_{BW} = x_2$, $\gamma_{WB} = x_3$ and $\gamma_{WW} = x_4$ (in the sense of a statistical law). Obviously: $\Sigma_{i=1}^{4} x_i = 1$.

But one can analyse the results of the selection of two balls either according to the colour of the first ball (with the corresponding relative proportions γ_1 and γ_2) or according to the colour of the second ball (with the corresponding relative proportions γ_1' and γ_2').

We have:

$$\gamma_1 = x_1 + x_2,$$
$$\gamma_2 = x_3 + x_4,$$
$$\gamma_1' = x_2 + x_4,$$
$$\gamma_2' = x_1 + x_3.$$

But the inverse problem could be posed: Find out $\{x_i\}$ when $i = \overline{1, 4}$, and γ_1, γ_2, γ_1', γ_2' are given.

The following system of linear equations can be written:

$$
\begin{aligned}
x_1 &+ x_2 & & & &= \gamma_1, \\
 & & x_3 &+ x_4 &= 1 - \gamma_1, \\
 & x_2 & &+ x_4 &= \gamma_1', \\
x_1 & &+ x_3 & &= 1 - \gamma_1'.
\end{aligned}
$$

Since $x_4 = 1 - x_1 - x_2 - x_3$, there are only three unknowns ($n = 3$). There are only two known values ($m = 2$) since $\gamma_2 = 1 - \gamma_1$, and $\gamma'_2 = 1 - \gamma'_1$.

The system can thus be written:

$$(A) \quad \begin{aligned} x_1 + x_2 \qquad &= \gamma_1, \\ x_1 \qquad + x_3 &= \gamma'_2. \end{aligned}$$

Two questions may be asked:
(a) is the system A compatible?
(b) is the system A determined?

Let us introduce now two matrices:

$$\|A\| = \begin{Vmatrix} 1 & 1 & 0 \\ 1 & 0 & 1 \end{Vmatrix} \quad \text{of rank 2}$$

and

$$\|B\| = \begin{Vmatrix} 1 & 1 & 0 & \gamma_1 \\ 1 & P & 1 & \gamma_2 \end{Vmatrix} \quad \text{of rank 2.}$$

The system A is compatible because

$$\text{rank } \|A\| = \text{rank } \|B\|.$$

But $r = m$ and $r < n$.

Thus the system A is not determined. That is the reason why the rule of conditional probabilities cannot be said to master the situation. The given answer of the rule is only one element of the infinite set of allowed solutions.

APPENDIX II.[2] IS BERNOULLI'S DISTRIBUTION
ALSO THE LIMIT DISTRIBUTION?

n white balls and m black balls are inside a box. We take at random k balls out of this box ($k < n + m$). Which will be the relative part of the subset of the results for which there are r black balls among the k balls selected ($r \leqslant m$)? Usually this subset can be divided into $k + 1$ classes according to the value of k.

Number of balls

Black	0	1	2	3	$\cdots$	$k-1$	k
White	k	$k-1$	$k-2$	$k-3$	$\cdots$	1	0

The relative part of each of these classes $B_k(r)$ is calculated:

Number of black balls	Relative part $B_k(r)$
0	$C_n^k C_m^0$
1	$C_n^{k-1} C_m^1$
2	$C_n^{k-2} C_m^2$
. . . .	. . . .
r	$C_n^{k-r} C_m^r$
. . . .	. . . .
k	$C_n^0 C_m^k$

$$B_m(r) = \frac{C_n^{k-r} C_m^r}{\sum\limits_{i=0}^{k} C_n^{k-i} C_m^i} = \frac{C_n^{k-r} C_m^r}{C_{n+m}^k}$$

$$= \frac{n!\, m!\, k!\, (n+m-k)!}{(k-r)!\, (n-k+r)!\, (m-r)!\, r!\, (m+n)!}.$$

By using Stirling's formula:

$$S! = \sqrt{2\pi S}\, S^s\, e^{-s}\, e^{\theta_s},$$

in which $|\theta_s| \leqslant 1/(12\,s)$; it can be shown that:

$$B_m(r) = C_k^r \frac{\sqrt{nm(n+m-k)}}{\sqrt{(n+m)(n-k+r)(m-r)}} \left(\frac{n}{n+m}\right)^{k-r}$$

$$\times \left(\frac{m}{n+m}\right)^r \left(\frac{m}{m-r}\right)^{m-r} \left(\frac{n}{n-k+r}\right)^{n-k+r}$$

$$\times \left(\frac{n+m-k}{n+m}\right)^{n+m-k} e^{\theta},$$

in which

$$\theta \leqslant \frac{1}{12}\left(\frac{1}{n} + \frac{1}{m} + \frac{1}{m+n-k} - \frac{1}{n+m} - \frac{1}{m-r} - \frac{1}{n-k+r}\right).$$

n and m are supposed to go to infinity in such a way that $n/(n+m)$ tends towards b and $m/(n+m)$ tends towards a, a and b being real numbers. If $n+m$ tends towards infinity, the root term tends towards 1, but θ tends towards 0. This means that:

$$\lim B_m(r) = C_k^r a^r b^{k-r},$$

$$m + n \to \infty.$$

In other words, we obtain Bernoulli's distribution which is in fact the limit distribution.

APPENDIX III. ANALYSIS OF BERTRAND'S PROBLEMS
FROM THE STATISTICAL POINT OF VIEW

In 1888 Bertrand[12] set three problems about the geometrical probabilities which were known as Bertrand's paradoxes and which are known now to form but three different problems. In 1894 Poincaré[13] considered the fourth problem. Garwood and Hollroyd[14] set forth the fifth problem when dealing with the circulation of men and cars in a town.

What do these problems represent from the statistical point of view?

First Bertrand's Problem

Consider the set of the points on the circumference of some circle. Construct the Cartesian square product of these points. Each element of the square product set determines the chord of this circle (or a tangent if the components of the set are identical). This set can be divided into two subsets as follows; the chords that are shorter than the side of the inscribed regular triangle, form subset 1 and the others subset 2. The problem is to find out the relative proportion of the subset 1. One can fix a point on the circumference and look at the set of chords that correspond to the set of the second points of the chords. From the point of view of symmetry (maybe not a very strict one) we may conclude that the relative proportions of the subsets of these new chords will be the same, i.e. that these sets of chords are similar from the statistical point of view.

Second Bertrand's Problem

Let us consider a set of points on the fixed diameter of a given circle. We construct the set of the chords that are perpendicular to this diameter. It can be considered as a set of the chords that are made by a point at infinity and by the set of the points of half the circumference. We ask the same question for this set as we asked previously.

The statistical approach allows one to generalize these two problems of Bertrand without any difficulty.

Let us consider the set of the chords to a circle radius 1 made by the set of straight lines (AP).

When the length (L) of OP equals 1, we have the first problem; when P is located at the infinity, we have the second problem. We shall consider the general case $0 < L < \infty$ (Figures 1 and 2).

Let us suppose now that the angle APO is always distributed uniformly. Since the length of the line AB is $e = 2\sqrt{1 - L^2 \sin^2 \alpha}$ $(L < 1)$ the density of the distribution is

$$\omega(e) = \frac{2}{\pi} \frac{e}{\sqrt{4L^2 - 4 + e^2}\sqrt{4 - e^2}}.$$

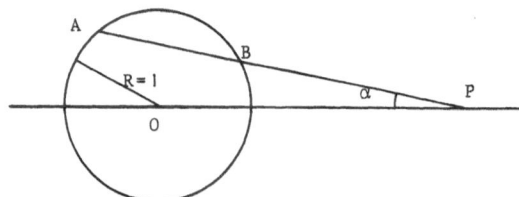

Fig. 1.

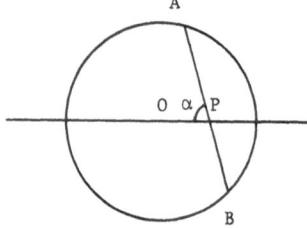

Fig. 2.

When $L \to 1$,

$$\omega(e) \to \frac{2}{\pi \sqrt{4-e^2}}.$$

If $1 < L < \infty$,

$$\omega^*(e) = \frac{1}{\pi} \arcsin \frac{e^2-4+2L^2}{2L^2} + \frac{1}{2}.$$

When $L \to 1$,

$$\omega^*(e) \to \frac{2}{\pi \sqrt{4-e^2}},$$

but when $L \to \infty$,

$$\omega^*(e) \to \frac{e}{2\sqrt{4-e^2}}.$$

The corresponding distribution functions will be:

$$F(e) = \frac{1}{\pi} \arcsin \frac{e^2-4+2L^2}{2L^2} + \frac{1}{2},$$

$$F^*(e) = \frac{1}{2 \arcsin (L^{-1})} \left[\arcsin \frac{e^2-4+2L^2}{2L^2} - \arcsin \frac{L^2-2}{L^2} \right].$$

When $L \to 1$,

$$F_1(e) \to \frac{1}{\pi} \arcsin \frac{e^2-2}{2} + \frac{1}{2}.$$

The solution of the first Bertrand's problem will be:

$$\phi(1) = 1 - F_1(\sqrt{3}) = 1 - \frac{1}{\pi} \arcsin \left(\frac{3-2}{2} \right) - \frac{1}{2} = \frac{1}{3}.$$

When $0 < L < 1$,

$$\phi(L) = 1 - F(\sqrt{3}) \frac{1}{2} - \frac{1}{\pi} \arcsin \frac{2L^2-1}{2L^2}.$$

When $L \to 1$,

$$F^*(e) \to F_1(e).$$

When $L \to \infty$,

$$F^*(e) \to 1 - \frac{\sqrt{4-e^2}}{2}.$$

In this case,

$$\phi^*(\infty) = 1 - F^*(\sqrt{3}) = 1 - \frac{\sqrt{4-3}}{2} = \frac{1}{2}.$$

In the most general case: $1 < L < \infty$,

$$\phi^*(L) = 1 - F^*(\sqrt{3})$$

$$= \frac{1}{2 \arcsin (L^{-1})}\left[\arcsin \frac{2L^2-1}{2L^2} - \arcsin \frac{L^2-2}{2} \right].$$

The diagram of $\phi(L)$ and $\phi^*(L)$ is presented on Figure 3.

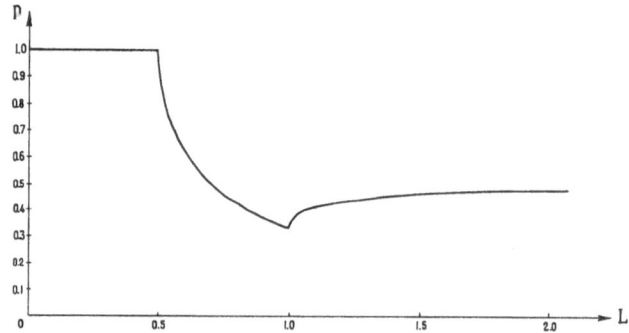

Fig. 3. The relative part of the set of chords the lengths of which are larger than the length of the side of a regular inscribed triangle.

APPENDIX IV. ON THE EXPLANATION OF THE EMPIRICAL LAWS OF BROWNIAN MOTION AND THE DERIVATION OF EINSTEIN'S EQUATION FOR UNIDIMENSIONAL WALKS OF SETS

1. Experience shows that:

(a) when the mass of a Brownian particle gets smaller, its motion becomes more agitated,

(b) when the temperature of the ambient liquid rises, the motion becomes also more agitated.

These qualitative laws (Ch. Wiener, J. Delsaux, J. Thirion, M. Gouy, S. Exner and F. Exner) are the consequences of the observation of the phenomenon. But they can be explained within the frame of the molecular theory.

The Explanation of (a)

A particle at rest in suspension in a liquid is supposed to be compared to a sphere of mass M and a molecule to be compared to a sphere of mass $m(m \ll M)$. The particle can collide only with one molecule at a time, for instance a molecule number j, whose speed at the considered moment is v_j. The collisions are supposed to be absolutely elastic and central. After the collision with the sphere the speed of the large particle will be $v_j = 2mv_j/(M + m)$. The displacement of a Brownian particle can be observed if v_j is larger than a minimal value v_0. Thus the displacement can be observed if $v_j > v_{j_0} = 0.5v_0((M + m) + 1)$, in which v_{j_0} is the minimal value that can produce an observable shift of a Brownian particle. If M increases, v_{j_0} increases too. Observable displacements will be rarer if M is greater. On the contrary they will be more frequent if M is smaller.

The Explanation of (b)

Let us now consider Brownian motion in a gas. From the point of view of the kinetic theory, the cause of Brownian motion is the same as in the liquid. The temperature T of the gas we take as given. The total number of gas molecules is N; the number of molecules with speed less than v_j is N_1. If $T_2 > T_1$, the corresponding number of molecules is N_2, greater than N_1. Thus when the temperature of the gas rises, Brownian motion will be more agitated. The same will result when the mass of Brownian particles are taken to be smaller.

2. By using the statistical concept of the random walk, we may obtain Einstein's law in an elementary way. There are N parallel lines and N particles which can move either forward or backward during a unit time interval. The length of each single displacement is also 1 unit of length. This corresponds to the collisions of Brownian particles with only one molecule at any given instant. At time $t = 0$, all the particles have zero spatial coordinates. We say that these particles execute the

same motion if their laws of motion coincide with one another. For $t = n$, there are 2^n different possible paths and the particles can have $n + 1$ different positions. The possible paths are numbered from 1 to 2^n and the whole set of N particles $(N \geqslant 2^n)$ is arranged in groups according to their paths.

When i is the path index, N_i is the number of the particles that execute path i.

Let us suppose that $\beta_n(i) = N_i/N$ ($\beta_n(i)$ is a relative fraction of the subset characterized by the path number i). Obviously: $\Sigma_{i=1}^{2^n} \beta_n(i) = 1$. But, on the other hand, the set of N particles can be arranged in groups according to their positions at time $t = n$.

The set of the possible coordinates at the instant $t = n$ is given by $x(k) = n - 2k$, in which $k = \overline{0, n}$, i.e. there are only $n + 1$ subsets. The relative fraction of a subset corresponding to the position $x(k)$ is $\gamma_n(k)$.

Obviously

$$\sum_{n=0}^{n} \gamma_n(k) = 1 \quad \text{and} \quad \gamma_n(k) = \sum_{i=C_n^{k-1}+1}^{C_n^k} \gamma_n(i),$$

in which C_n^m is the binominal coefficient.

In theory, the distributions $\beta_n(i)$ and $\gamma_n(k)$ can be found out from experiment, but actually it is impossible. A supplementary hypothesis is needed. We assume that $\beta_n(i) = 2^{-n}$, that is to say that the distribution of paths is uniform, in which a case $\gamma_n(k) = 2^{-n} C_n^k$.

Then we introduce the notion of mean position of the particles, $\langle x \rangle$. From the point of view of mechanics, $\langle x \rangle$ is only a position of the centre of mass of a system of particles of equal mass.

In our case

$$\langle x \rangle = \sum_{k=0}^{n} (n - 2k)\gamma_{nk} = n + (n - 2)n + \cdots + (n - k)C_n^k + \cdots$$
$$+ (-n + k)C_n^{k-n} + \cdots + (-n + 2)n - n = 0.$$

Thus the centre of mass of the particles remains at rest.

We can also calculate $\langle x^2 \rangle$ which is proportional to the moment of inertia of the particles relative to an axis of rotation passing through the origins of the N straight lines.

$$\langle x^2 \rangle = \sum_{k=0}^{n} (n - 2k)^2 \frac{C_n^k}{2^n} = \frac{1}{2^n} \sum_{k=0}^{n} (n^2 - 4nk + 4k^2) C_n^k$$

$$= \frac{n^2}{2^n} \sum_{k=0}^{n} C_n^k - \frac{4n}{2^n} \sum_{k=0}^{n} kC_n^k + \frac{4}{2^n} \sum_{k=0}^{n} k^2 C_n^k.$$

For $\displaystyle\sum_{k=0}^{n} kC_n^k = n2^{n-1}$,

$$\sum_{k=0}^{n} k^2 C_n^k = 2^{n-2} n(n+1),$$

$$\langle x^2 \rangle = \frac{n^2 \cdot 2^n}{2^n} - \frac{4n^2 \cdot 2^{n-1}}{2^n} - \frac{4 \cdot 2^{n-2}(n+1)n}{2^n}$$

$$= n^2 - 2n^2 + n^2 + n = n.$$

This is simply Einstein's law when $t = n$ and with the coefficient of diffusion $D = 0.5$. D could be introduced arbitrarily by varying the intervals of time and the distances covered during one unit of time.

Let us now try to find out what happens in the case of ternary collisions, i.e. when the particle is subjected to impacts with two other molecules at the same moment.

At the instant t, there are $2n + 1$ possible positions

$$x(k) = n - k + 1, \qquad \text{or} \quad k = \overline{1, 2n + 1}.$$

Similarly there are 3^n different paths.

If we suppose that the uniformity of the distribution with respect to the motion is possible, we obtain $\gamma_n(k) = 3^{-n} C_3(k, n)$, in which $C_3(k, n)$ are numbers of the nth line of a 3-arithmetical triangle. (See for example [15].) Thus

$$\langle x \rangle = \sum_{k=1}^{2n+1} (n - k + 1) \frac{C_3(k, n)}{3^n}$$

$$= \frac{n+1}{3^n} \sum_{k=1}^{2n+1} C_3(k, n) - \frac{1}{3^n} \sum_{k=1}^{2n+1} kC_3(k, n)$$

$$= \frac{n+1}{3^n} S_{30}(n) - \frac{1}{3^n} S_{31}(n),$$

where

$$S_{30} = \sum_{k=1}^{2n+1} C_3(k, n) = 3^n,$$

equal to the sum of the numbers of the nth line of the 3-arithmetical triangle. (See [15], pp. 131–133.)

$$S_{31} = \sum_{k=1}^{2n+1} kC_3(k, n) = (n + 1)3^n.$$

Thus

$$\langle x \rangle = \frac{n+1}{3^n}3^n - \frac{1}{3^n}(n + 1)3^n = 0.$$

The mean square of the position $\langle x^2 \rangle$ is equal to

$$\sum_{k=1}^{2n+1} (n - k + 1)^2 \frac{C_3(k, n)}{3^n}$$

$$= \frac{(n + 1)^2}{3^n}S_{30}(n) - \frac{2(n + 1)}{3^n}S_{31}(n) + \frac{1}{3^n}S_{32}(n),$$

in which

$$S_{32}(n) = \sum_{k=1}^{2n+1} k^2 C^3(k, n) = (3n^2 + 8n + 3)3^n.$$

Thus

$$\langle x \rangle^2 = \frac{(n + 1)^2}{3^n}3^n - \frac{2(n + 1)^2}{3^n}3^n + \frac{3n^2 + 8n + 3}{3}$$

$$= n^2 - 2n - 1 + \frac{3n^2 + 8n + 3}{3} = \frac{2n}{3}.$$

In this case, the coefficient of diffusion becomes $\frac{1}{3}$ since ternary collisions reduce the mean velocity of the particles which go away from the origins of the axes.

Lomonosov University
Moscow

NOTES

* Translated from the French by Yves Paty.
[1] The author apologizes for renaming the fundamental statistical terms but this is only done to facilitate the strict methodological definition of statistical law.
[2] Appendices II, III, and IV are the exposition of results obtained in collaboration with

L. V. Matyushkin, former undergraduate of Physics Department, Lomonosov State University, Moscow, U.S.S.R.

BIBLIOGRAPHY

[1] Ballentine, L. E., *Rev. Mod. Physics* **42** (1970), 358–381.
[2] Born, M., *Physik* **37** (1926), 863–867; **38** (1926), 803.
[3] Born, M., *Nature* (Great Britain) **119** (1927), 354–357; *Naturwissenschaften* **15** (1927), 238–242.
[4] Elsevier, 1965.
[5] Leipzig-Wien, 1931.
[6] *Math. Z.* **5** (1919), 52–99.
[7] See, for example, Ville, J., *Étude critique de la notion de collectif*, Gauthier-Villars, Paris, 1939.
[8] There are some references devoted to the taxonomical problems in particular branches of sciences and humanities: Sokal, R. R. and Sneath, P. M. A., *Principles of Numerical Taxonomy*, N.Y., Toronto, 1963 (There is a more recent edition); Gardin, J. C., *Études archéologiques* (ed. by P. Courbin), Paris, 1963, pp. 132–150; *American Antiquity* **32** (1967), 13–30; *Archéologie et Calculateurs*, Problèmes sémiologiques et mathématiques, C.N.R.S., Paris, 1970; Vega, W. Fernandez de la, *Revue Française de sociologie* **8** (1967), 506–520; Lerman, I. C., *Revue de statistique appliquée* **21** (1973), 23–49; Kroeber, A. L., *Language* **36**, No. 1, part I (1960), 1–22.
[9] Czekanowski, J., *Zarys metod statystacznych w sastosowanium do, antropologii*, Warszawa, 1913; *Zarys antropologii Polski*, Lwow, 1930.
[10] *Przeglad antropologiczny* **17** (1951), 193–211; Florek, K., Lukaszewicz, J., Perkal, J., Steinhaus, H., and Zubrzycki, S., *Colloquium Mathematicum* **2** (1951), 282–285; Mikiewicz, J., *Zastosowania matematyki* **7** (1963), 40.
[11] Kemeny, J. G., Snell, J. L., and Thompson, G. L., *Introduction to Finite Mathematics*, Prentice-Hall, New York, 1957.
[12] Bertrand, J., *Traité de Calcul des probabilités*, Paris, 1888, pp. 4–5.
[13] Poincaré, H., *Calcul des probabilités*, Paris, 1894; 2nd ed., Paris, 1912, pp. 118–130.
[14] Garwood, F. and Holroyd, E. M., *Mathematical Gazette* **50**, 2nd ed. (1966), 283–286.
[15] Vilenkin, N. Ya., *Kombinatorika*, Moskva, Nauka (1969) (in Russian), pp. 129–130.

M. MUGUR-SCHÄCHTER

THE QUANTUM MECHANICAL ONE-SYSTEM FORMALISM, JOINT PROBABILITIES AND LOCALITY

"Il ne faut pas que l'esprit
s'arrête avec les yeux, car la vue
de l'esprit a bien plus d'étendue
que la vue du corps".

Malebranche

R. Magritte

1. INTRODUCTION

Professor Wigner[1] has proved a theorem which is believed to establish the impossibility of associating with any state vector a joint probability of the position and momentum variables. In this work we study this important theorem and we show that in fact it does not rule out the joint probability concept, but that instead it leads to a locality problem inside the one-system formalism of quantum mechanics, similar in certain respects to the problem formulated by Bell[2] inside the two-systems formalism of quantum mechanics.

The analyses which we carry out draw attention to the super-position states with non-connected support, raising doubt concerning the truth of certain quantum mechanical predictions for such states.

J. Leite Lopes and M. Paty (eds.), Quantum Mechanics, a Half Century Later, 107–146. All Rights Reserved

2. STUDY OF WIGNER'S THEOREM ON JOINT PROBABILITIES

2.1. *Wigner's Demonstration*

We start by reproducing Wigner's demonstration. This will be done in detail, in order to facilitate any eventual comparison.

Given a one-system wave function $\psi(q)$ (in one-dimensional notation), Wigner studies a joint function $P(q, p)$ of the positional variable q and the momentum variable p, on which he imposes the following conditions:

(a) that it be a 'hermitian form of $\psi(q)$', i.e.

(1) $P(q, p) = (\psi, M(q, p)\psi),$

where M is a self-adjoint operator depending on p and q, and

(b) that $P(q, p)$, if integrated over p, give the proper probabilities for the values of q, as

(2a) $\int P(q, p)\, dp = |\psi(q)|^2,$

and, if integrated over q, give the proper probabilities for the momentum, as:

(2b) $\int P(q, p)\, dq = (2\pi\hbar)^{-1} \left| \int \psi(q)\, e^{-ipq/\hbar}\, dq \right|^2.$

The condition (b) admits the somewhat milder substitute that $P(q, p)$ should give the proper expectation value for all operators which are sums of a function of p and a function of q, as

(2) $\iint P(q, p)(f(p) + g(q))\, dq\, dp = \left(\psi, \left(f\left(\frac{\hbar}{i}\frac{\partial}{\partial q}\right) + g(q) \right) \psi \right).$

A third 'very natural' condition on $P(q, p)$ would be that it is non-negative for all values of q and p:

(3) $P(q, p) \geq 0.$

But Wigner demonstrates that the conditions (a) and (b) are incompatible with (3). This is realized by showing that the assumption that a $P(q, p)$ satisfying all three conditions (a), (b) and (3) can be defined for *every* ψ, leads to a contradiction.

The contradiction is obtained for wave functions $\psi(q)$ of a particular form, namely for ψ which are linear combinations $(a\psi_1 + b\psi_2)$ of any two fixed functions such that ψ_1 vanishes for all q for which ψ_2 is non-null, and vice versa. Wigner starts with the following lemmas:

LEMMA 1. If $\psi(q)$ vanishes in an interval I, and if $g(q)$ is zero outside this interval and nowhere negative therein, one has for the P corresponding to the $\psi(q)$ above:

(4) $$\int P(q, p)g(q)\, dq = 0,$$

for all p (except for a set of measure zero).

This follows from (2) with $f = 0$: the integral of (4) with respect to p vanishes because the right side of (2) vanishes

(4a) $$\iint P(q, p)g(q)\, dp\, dq = (\psi, g(q)\psi) = 0.$$

However, the integrand with respect to p, that is the left side of (4), is non-negative for the g postulated, as long as (3) holds for P. It follows then that the integrand with respect to p must vanish except for a set of p of measure zero, q.e.d.

Furthermore, (4) is valid for every function $g(q)$ which satisfies the conditions of Lemma 1. It can then be concluded in a similar way that:

LEMMA 2. If $\psi(q)$ vanishes in an interval I, the corresponding $P(q, p)$ vanishes for all values of q in that interval (except for a set of measure zero).

Wigner's demonstration then continues as follows:

Let us consider two functions $\psi_1(q)$ and $\psi_2(q)$ which vanish outside of two nonoverlapping intervals I_1 and I_2 respectively. Because of (1), the distribution function $P_{ab}(q, p)$ which corresponds to $\psi = a\psi_1 + b\psi_2$ will have the form:

(5) $$P_{ab}(q, p) = |a|^2 P_1 + a^*b P_{12} + ab^* P_{21} + |b|^2 P_2.$$

Setting $b = 0$, we note that P_1 is the distribution function for ψ_1, and similarly, setting $a = 0$, P_2 is the distribution function for ψ_2. Let us consider (5) for the q outside the interval I_1. Since (according to Lemma 2) P_1 vanishes almost everywhere for such q, the distribution function (5) cannot be positive for all a and b unless both P_{12} and P_{21} vanish if q is outside I_1 (except for a set of measure zero in q and p). A similar conclusion can be drawn when q is outside I_2. Hence, we have instead of (5), almost everywhere,

(6) $P_{ab}(q, p) = |a|^2 P_1(q, p) + |b|^2 P_2(q, p)$.

This means that the distribution function P_{ab} is almost everywhere independent of the complex phase of a/b. But this is impossible if P_{ab} is to give the proper momentum distribution for $\psi = a\psi_1 + b\psi_2$, i.e. is to satisfy (2b). Indeed, let us denote the Fourier transforms of $\psi_1(q)$ and $\psi_2(q)$ by $\Phi_1(p)$ and $\Phi_2(p)$. Equation (2b) then reads

(7) $|a|^2 \int P_1(q, p)\, dq + |b|^2 \int P_2(q, p)\, dq$

$$= |a|^2 |\Phi_1(p)|^2 + |b|^2 |\Phi_2(p)|^2 + 2Re\, ab^* \Phi_1(p)\Phi_2^*(p).$$

Since this must be valid for all a and b, it requires identically in p:

(7a) $\Phi_1(p) \cdot \Phi_2^*(p) = 0$.

But this is impossible, since $\Phi_1(p)$ and $\Phi_2(p)$, being Fourier transforms of functions restricted to finite intervals, are analytic functions (in fact, entire functions) of their arguments, and cannot vanish over any finite interval.

Professor Wigner formulates the result of his demonstration in the following terms (p. 28):

"no non-negative distribution function can fulfil both postulates (*a*) and (*b*)".

2.2. *Bearing of Wigner's Theorem*

Preliminaries

There seems to be a tendency to interpret Wigner's theorem as the expression of an absolute impossibility of a joint probability of the position and momentum associable to the quantum mechanical state vectors. Such a tendency betrays the real conceptual situation.

Quite generally a *demonstrated* absolute impossibility is impossible: the framework inside which an impossibility is demonstrated ineluctably restricts its bearing. Some of these restrictions cannot be suppressed without disintegrating the studied problem, but some of them might not be essential to the definition of the problem, or even might vitiate it. Obviously only an explicit examination of the logical relativities of a proposition to the framework of its proof can show which restrictions can or must be dropped.

Furthermore the bearing of a theorem is relative also to the inner structure of the proof (via one counter example, or directly for the whole class considered).

We shall now examine the various logical relativities of Wigner's theorem, which define its bearing.

Framework of the proof
The framework consists of the postulates: (a) (hermitian forms defined by (1)), (b) (the two marginal conditions (2) for any ψ), and the non-negativity condition (3). The assumptions of non-negativity and of hermiticity are entailed by the significance of a probability required for the distribution $P(q, p)$, hence they cannot be dropped without disintegrating the very problem chosen for examination, which consists precisely in the possibility of a *probability* distribution $P(q, p)$. Thus eventual unnecessary restrictions can be implied only in Definition (1) and/or in Postulate (b).

DEFINITION (1). Definition (1) is not the most general one conceivable. The distribution operator M is required self-adjoint and dependent exclusively on q and p. The second requirement entails for M independence on ψ, and this entails $P(q, p)$ as a *sesquilinear* form of ψ. Now the functional $P(q, p)$ is researched such as to accept the significance of a probability. Then the concept of a probability requires by its definition the reality of $P(q, p)$ so that $P(q, p)$ must be indeed a hermitian form of ψ: the condition that M be self-adjoint cannot be dropped. But the independence of M on ψ is not imposed via the probabilistic significance desired for $P(q, p)$, so that in the examined context it is an arbitrary a priori restriction. We shall now show that:

PROPOSITION. In absence of the arbitrary restriction to a sesquilinear form for $P(q, p)$, Wigner's demonstration cannot be realized.

Proof. Instead of (1) we start out with the most general definition a priori conceivable for a joint probability distribution of q

and p, namely

(1)' $P(q, p) = (\psi, M(q, p, \psi)\psi)$,

where the distribution operator $M(q, p, \psi)$ is self-adjoint and depends on q, p and ψ. All the other assumptions introduced by Wigner are left unchanged. We introduce the notations: ψ_{ab} is a state vector $a\psi_1 + b\psi_2$ where the supports of ψ_1 and ψ_2 are disjoint; P'_{ab}, P'_1, P'_2 are respectively the distributions obtained for ψ_{ab}, ψ_1 and ψ_2 by use of Definition (1)'; P'_{12} and P'_{21} are respectively the analogs of P_{12} and P_{21} from (5) obtained by use of (1)'. With these notations the expression of the joint distribution for ψ_{ab} yielded by Definition (1)' is

(5)' $P'_{ab}(q, p) = |a|^2(\psi_1, M(q, p, \psi_{ab})\psi_1) + a^*bP'_{12} + ab^*P'_{21}$
$$+ |b|^2(\psi_2, M(q, p, \psi_{ab})\psi_2).$$

In Wigner's expression (5), the factor of $|a|^2$ in the first term and the factor of $|b|^2$ in the last term identify respectively with the distribution P_1 yielded for ψ_1 by Definition (1) and with the distribution P_2 yielded for ψ_2 by Definition (1). The sequel of Wigner's proof is directly founded on this fact and on Lemma 2, as it can be verified by inspection. But this fact is *not* reproduced in Expression (5)'. Now this is so precisely because of the dependence on ψ of the distribution operator M from (1)', which introduces ψ_{ab} in the argument of M, instead of, respectively, ψ_1 in the factor $|a|^2$ and ψ_2 in the factor of $|b|^2$. For this reason – even though Lemma 2 continues to hold in the assumed context – Wigner's proof can no more be reproduced with the nonsesquilinear definition (1)', q.e.d.

If not Wigner's proof, then Wigner's conclusion might be generalizable – by some other proof – to any definition of a joint probability subject to both marginal conditions (2). But in fact this cannot be done either, as a well-known example suffices to show: the 'trivial' or 'correlation-free' distribution $|\psi(q)|^2|\Phi(p)|^2$ (where Φ is the Fourier transform of ψ) is a non-negative hermitian and *non*-sesquilinear form of ψ defined for any ψ and which fulfils both marginal conditions (2). Therefore it can be concluded that Wigner's theorem has no bearing on a non-void class of joint probabilities a priori possible. On mathematical grounds (considerations of continuity) it seems probable that this class is not reduced to the trivial distribution alone. It cannot be decided whether this class contains or not 'interesting' members, as long as the structure of all the conditions to be imposed upon a joint probability (time evolution, mean conditions, correspondence rules between functions and operators, etc. . . .) has not yet been thoroughly defined and studied as an organic whole. The attempts made up to now in this direction are not numerous and – as far as we know – none of them is both complete and guided by an explicit and coherent system of *physical* criteria for the choice of the mathematical conditions.

Postulate (*b*). Let us now examine the two marginal conditions (2). In a first approach we admit the truth of the quantum mechanical predictions expressed by the second members of (2), for any ψ. In a second approach we question this truth for the particular states described by vectors ψ_{ab}.

First stage: The truth of the predictions from the second members of Relation (2) being a priori posed for any ψ, the conditions of consistency with quantum mechanics expressed by use of the first members of (2) are not the most general ones conceivable. They are in fact very restrictive, requiring the *observability* of the integrated distributions $P(q) = \int P(q, p) \, \mathrm{d}p$, $P(p) = \int P(q, p) \, \mathrm{d}q$ (even though not necessarily of the values q, p also). The joint probabilities $P(q, p)$ subjected to less restrictive conditions of consistency escape Wigner's theorem.

Second stage: An exhaustive examination of the logical relativities of Wigner's theorem obliges us to raise finally also the question of the truth of the second members of both conditions (2) *for the particular state vectors ψ_{ab} with non-connected support*. Indeed Wigner's theorem being based on a counter-example proved for the mentioned states, the theorem would remain without foundation if for these particular states the right-hand members from (2) were not both true. This question of truth, even though brought in merely by logical considerations, seems less irrelevant from the physicist's point of view when it is realized that probably the momentum distribution in a state ψ_{ab} with non-connected support has never been measured, so that the 'existence' of an interference term is so far a purely formal fact; not even the assertion of *measurability* of the momentum 'observable' seems to have an obvious operational meaning, neither for such states in particular, nor in general (more detailed remarks can be found on pp. 132, 134, 135).

Inner structure of the proof

Wigner's theorem is demonstrated by producing a counterexample to the initial assumptions, which holds for the state vectors of the particular type $\psi_{ab} = a\psi_1 + b\psi_2$ where the supports of ψ_1 and ψ_2 are disjoint. Even though via this counterexample a *general* impossibility (for any ψ) is established indeed, this impossibility, nevertheless, has no bearing on the sub-class of state vectors of a type different from ψ_{ab}, which contains the major part of the state vectors coming usually into consideration: the theorem leaves open the question whether yes or not for the state vectors $\psi \neq \psi_{ab}$ a non-negative form (1) can fulfil both marginal conditions (2). In certain contexts this question might appear as non-trivial from the physicist's point of view (if, for instance, the quantum mechanical predictions for the momentum in states ψ_{ab} were false).

Conclusion

The preceding analysis shows that Wigner's proof does not exclude the possibility of any non-negative joint distribution function of the position and momentum variables associated with the quantum mechanical state vectors.

Notwithstanding this conclusion we believe that Wigner's proof has an outstanding heuristic interest. Indeed, once an analyzed knowledge has been obtained concerning its structure and its bearing, this proof suggests developments which disclose questions of a fundamental conceptual importance. The remainder of this article is devoted to these developments.

3. SUPERPOSITION STATES WITH NON-CONNECTED SUPPORT AND NON-LOCALITY OF THE ONE-SYSTEM FORMALISM OF QUANTUM MECHANICS

3.1. *The Problem*

The counterexample on which Wigner's theorem is based possesses characteristics which suggest the possibility of a problem of locality implicit in the one-system formalism of quantum mechanics. Indeed, the state vector directly concerned by the proof is a superposition vector $\psi_{ab} = a\psi_1 + b\psi_2$ with non-connected support. The distributions of the position and of the momentum predicted by quantum mechanics for such a state are respectively

(8) $\qquad |\psi_{ab}(q)|^2 = |a|^2|\psi_1(q)|^2 + |b|^2|\psi_2(q)|^2,$

and

(9) $\qquad |\Phi_{ab}(p)|^2 = |a|^2|\Phi_1(p)|^2 + |b|^2|\Phi_2(p)|^2 + 2Re\ ab^*\Phi_1(p)\Phi_2^*(p),$

where Φ_{ab}, Φ_1, Φ_2 are the Fourier transforms of, respectively, ψ_{ab}, ψ_1, ψ_2. Suppose now a joint probability $P(q, p)$ which fulfils the marginal conditions (2) for any ψ, hence in particular also for ψ_{ab} (by the analysis of Wigner's proof we know that such a joint probability, *if* it exists, cannot have a distribution operator independent of ψ). If the factor $P_{p/q}(p, q)$ of conditional probability of p given q is explicitly written, the marginal condition for q applied to ψ_{ab}, ψ_1 and ψ_2 leads (with obvious notations) to

(10) $\qquad P_{ab}(q, p) = P_{ab}(q)P_{ab, p/q}(q, p)$
$\qquad\qquad\qquad = |a|^2P_1(q)P_{ab, p/q}(q, p) + |b|^2P_2(q)P_{ab, p/q}(q, p).$

When we now examine (10) we are struck by the following aspect: For each given pair of values q_1, p_k one of the two terms of (10) is null, since either $q_1 \in I_1$ and then $q_1 \neq I_2$, or vice versa. Nevertheless when the conditional factor $P_{ab, p/q}(q, p)$ is tied to a value of the position variable belonging to I_1 we have in general

(11) $\qquad P_{ab, p/q}(q_1 \in I_1, p_k) \neq P_{1, p/q}(q_1 \in I_1, p_k),$

and when $P_{ab, p/q}(q, p)$ is tied to a value of q belonging to I_2 we have in general

(12) $\qquad P_{ab, p/q}(q_1 \in I_2, p_k) \neq P_{2, p/q}(q_1 \in I_2, p_k).$

This is so because (9) and the marginal condition for p applied to ψ_{ab}, ψ_1 and ψ_2 entail in general for $P_{ab}(p) = \int P_{ab}(q, p)\,dq$ that

(13) $\qquad P_{ab}(p) \neq P_1(p) + P_2(p).$

(Wigner's argument: the product $\Phi_1(p)\Phi_2^*(p)$ from (9) is not null identically in p for any ψ_1, ψ_2.) Thus when $P_{ab, p/q}(q, p)$ is tied to I_1 alone, its *value* is not determined only by ψ_1 with support I_1, it depends on the whole superposition $\psi_{ab} = a\psi_1 + b\psi_2$, and this is so notwithstanding the fact that the support I_2 of ψ_2 is separated from I_1 by an *arbitrary distance* (the symmetric proposition holds when

$P_{ab,\,p/q}(q, p)$ is tied to I_2 alone). This is a *mathematical* non-locality of the functional dependence on ψ_{ab} of the conditional probability $P_{ab,\,p/q}(q, p)$, emerging in the confrontation between the supposed joint probability $P_{ab}(q, p)$ and the topological characteristics of the support of ψ_{ab}. What Wigner's proof really shows is that *a sesquilinear definition of $P(q, p)$ cannot engender this mathematical non-locality, while the marginal conditions* (2) *do demand it for superposition states ψ_{ab} with a non-connected support.*

Now the mathematical non-locality specified above expresses exclusively spatial aspects of the confrontation between the concept of a joint probability and the non-connectedness of the support of ψ_{ab}. Therefore – as it stands – it has no established relation with some *physical* problem of 'locality' in the sense of the theory of relativity, where time plays an essential role. Furthermore this mathematical non-locality might vanish like a non-essential aspect when conditions of consistency less restrictive than (2) are required for $P(q, p)$ on the basis of some more analyzed physical criteria of relevance of a joint probability. The aim of this section is to show that in fact the mathematical non-locality perceived in the example from Wigner's proof is an essential aspect of any relevant joint probability $P(q, p)$ (and of any other probability distribution derived from a relevant $P(q, p)$) and that this formal non-locality does entail a problem of physical non-locality inside the one-system formalism of quantum mechanics.

The pursuit of this aim will draw attention on specificities of the superposition states which distinguish these states fundamentally from the mathematical decompositions permitted by the expansion postulate. Along this path we shall be led to the notion that the superposition principle – even though it materializes a mathematical possibility and even though it permitted to describe so accurately the wave-like aspects manifested by certain *position* distributions of microsystems – might nevertheless introduce inadequate predictions, either false or unverifiable, for the dynamical quantities which depend on the momentum and for the spin.

3.2. *Criterion for the Choice of Conditions of Consistency*

Before researching whether the mathematical non-locality discerned in the example from Wigner's proof entails or not a problem of physical non-locality, we shall first specify conditions of consistency with quantum mechanics such as they determine a joint probability concept $P(q, p)$ at the same time *minimally* restricted and 'relevant'. This of course requires criteria of relevance. We believe that the efficient criterion is that of relevance to the 'reduction problem', which is the core of the multiform and now more than fifty years old controversy on the significance of the quantum mechanical formalism. This problem is well-known: the quantum mechanical formalism yields only a statistical prediction concerning the outcome of one individual act of measurement, while this act brings forth a unique well-defined result thereby 'reducing' the predicted spectrum to a certain certitude. The main purpose of those who desire a hidden variables substitute to quantum mechanics is to obtain a 'deterministic' solution for the reduction problem. Such a solution is researched along the following lines. It is postulated that the studied system possesses, independently of observation, certain intrinsic properties statistically describable by a virtual distribution of values of an appropriate group of hidden parameters (hidden to quantum mechanics but not necessarily also to observation). For one given system, at any given time, only one of all the possible groups of values for this group of hidden parameters is conceived to be realized. Each measurable 'quantity w of a system' is conceived as related with a corresponding function h_w of the hidden parameters. An individual act of measurement of w is conceived as a process of interaction between the system and a w-measurement device, which act induces into a deterministic evolution the unique but unknown value $h_{i,w}$ possessed by h_w at the initial moment of this act of measurement. The unique observed value w_j brought forth by the act of measurement can thus be considered to emerge as an observable result of the system-device interaction, deterministically connected with the unique preexisting initial value $h_{i,w}$ via the interaction evolution. It has to be stressed however that the existence of a deterministic connection between each observed w_j with *one* value $h_{i,w}$ does not entail a one-to-one relation between the *values* $h_{i,w}$ and the values w_i; the assumption of such a one-to-one relation is obviously not essential for a deterministic solution of the reduction problem. Therefore it would be unnecessarily restrictive.

Since the main objective of the hidden variables attempts is to develop a deterministic solution to the reduction problem, we shall discard in what follows the conditions of consistency which engender joint probabilities a priori inadequate for the research of a deterministic solution to the reduction problem.

3.3. *Inadequacy of both Marginal Conditions* (2)

The marginal conditions $\int P(q, p)\,dp = P(q) = |\psi(q)|^2$ and $\int P(q, p)\,dq = P(p) = (2\pi\hbar)^{-1}|\int \psi(q)\,e^{-ipq/\hbar}\,dq|^2$ require the observability of both statistical distributions $P(q)$ and $P(p)$. This does not entail that the individual values of the variables q and p have to be also observable, nor does it fix the physical significance to be assigned to the symbols q, p.

If the possible significances of q, p are considered, it is immediately obvious that the significance of 'pure observables' (i.e. values of some observable entities for which the denominations of 'position' and 'momentum' are decreed, but which are defined *exclusively* by the specification of some experimental circumstances involving the system, and where these entities emerge) cannot be relevant to the reduction problem: the criterion of relevance to this problem requires a definition of q, p independent of observation. Discarding then the pure-observable significance and postulating for q, p a significance independent of observation, we shall now show that, whatever hypotheses are chosen concerning the observability of the individual values q, p, the marginal conditions (2) engender a joint probability $P(q, p)$ which is either unnecessarily restricted or self-contradictory.

The beable significance for q, p. Any property possessed by a system independently of observation has been called by Bell a beable property. We like this denomination and we adopt it. We shall now specify in detail the two important particular concepts, of a beable position and of a beable momentum.

Beable position. By definition this concept consists of the assumption of beable properties of the system which possess characteristics describable with the aid of the classical quantity position, i.e. which in any referential are, at any given time, non-negligible only inside a finite and relatively small spatial domain. Such an assumption is equivalent to a minimal *model* of the object named 'system'. However – by its minimality – this model does by no means entail the naïve atomistic, multitudinist hypothesis concerning the structure of the microreality; the finiteness and the smallness of the domain inside which the conceived beable position properties are 'confined', are only *relative* to some specified (and modifiable) degree of approximation chosen for the description of these properties, while their 'existence' is defined only with respect to some specified but arbitrary range of spatial dimensions characterizing the chosen scale of (imagined) observation. The concept of the object called system itself, to which a beable position is assigned, emerges only relatively to some choices of such approximations and of such a scale. Thus the notion that a beable position is possessed by what is named system has nothing absolute in it. In particular it leaves open the problems of separability of the systems and of locality of the phenomena in which they are involved.

Beable momentum. It is not impossible to conceive a beable position which does *not* perform a continuous dynamics, but which merely consists of a discontinuous juxtaposition of an uninterrupted succession of locations possessed by some properties of the system, in the sense specified above. But this sort of a beable position would reproduce the 'essentially probabilistic' features which a deterministic solution for the reduction problem attempts to remove. Such a beable significance for q in the argument of a joint probability $P(q, p)$ would therefore yield a concept irrelevant to the reduction problem, so that we discard it. If then a beable position which does perform a continuous dynamics is assumed, ipso facto some definite continuous time variation of this beable position is assumed. This – by definition – is what we call a beable momentum.

The beable individual kinematic relation: Thus the assumption of a continuously moving beable position of a system is interdependent with the assumption of a beable momentum of this system. These two united assumptions are *equivalent* to the assumption of the descriptive relevance of a position variable q and a corresponding momentum variable p, *tied* to one another by the individual kinematic relation (in one dimensional writing)

$$(14) \qquad p = K \frac{dq}{dt},$$

where K is a factor of proportionality playing the role of an inertial mass. This individual relation is a non-trivial and important implication of the concept of a continuously moving beable position, because it entails statistical correlations and these can be found to be either compatible or incompatible with a given condition of consistency with quantum mechanics envisaged for a joint probability distribution of beable q, p.

Rejection of the requirement of both marginal conditions (2) *for beable q, p*: We consider two complementary hypotheses concerning the observability of beable values q, p, assumed to exist for a system: either not both these values are observable, or they are both observable. Either of these hypotheses leads to the rejection of the requirement of both marginal conditions (2). Indeed, we consider first an inobservable beable q or p. Then it can be rather trivially pointed out that:

PROPOSITION 1. The marginal conditions (2) entail an unnecessarily restricted statistical distribution of the values of an inobservable beable q or p.

Proof. Suppose that the value of the momentum beable is not observable for some given state of the studied system S. Let us then redenote this value p' in order to distinguish it from the observed value produced by an act of momentum measurement performed on S. Even though the individual values p' are not observable, the marginal condition (2b) requires that the statistical distribution $P(p')$ shall coincide with the observable quantum mechanical distribution $(2\pi\hbar)^{-1}|\int \psi(q) e^{-2\pi i p q / \hbar} dq|^2$ of the values p (i.e. to each unknown value p' corresponds one observed value p which arises statistically the same number of times). This, however, is an unnecessary restriction on the relation permitted between values p' and values p: For ensuring at the same time consistency with quantum mechanics and relevance to the reduction problem it suffices to require that the observed values p alone have the quantum mechanical distribution and that, furthermore, each one observed value p be connected by the measurement interaction evolution, with *one* preexisting value p' (included in a hidden distribution $P(p')$ in general different from the observed one).

An analogous argument holds for q.

We consider now observable beables q, p. We shall show that

THEOREM 1. A joint probability distribution $P(q, p)$ of observable beables q, p, cannot fulfil both marginal conditions (2) for any state vector.

Proof. We produce an example: Consider the state vector

$$\psi(q) = \frac{1}{\sqrt{2}}\,\phi_{\mathbf{p}_1}(q) + \frac{1}{\sqrt{2}}\,\phi_{\mathbf{p}_2}(q),$$

where $\phi_{\mathbf{p}_1}(q)$ and $\phi_{\mathbf{p}_2}(q)$ are eigendifferentials of the quantum mechanical observable momentum (vector), corresponding respectively to the eigenvalues $\mathbf{p}_1$ and $\mathbf{p}_2$ the directions of which make an angle $\alpha \neq 0$, the norms being equal and non-null ($|\mathbf{p}_1| = |\mathbf{p}_2|) \neq 0$. Since this state requires a two-dimensional description we refer it to two orthogonal axes ox, oz, the axis ox being chosen parallel to the bisectrix of α. The quantum mechanical position distribution $|\psi(x, z)|^2 = |\psi(\mathbf{q})|^2$ is then uniform along ox and periodic along oz; furthermore, this quantum mechanical distribution is *stationary*. We consider now a joint probability distribution $P(\mathbf{q}, \mathbf{p})$ associated with the chosen ψ and fulfilling both marginal conditions (2); $\mathbf{q}$ and $\mathbf{p}$ in the argument of $P(\mathbf{q}, \mathbf{p})$ are assumed to be observable beables. Then the beable character of $\mathbf{q}, \mathbf{p}$ entails that at each given time each instantaneous individual value of the momentum variable possesses a kinematic Definition (14) $\mathbf{p} = K(d\mathbf{q}/dt)$ according to which it is generated by the time variation of a corresponding joint $\mathbf{q}$. Via this kinematic definition and the hypothesis of observability of the individual $\mathbf{p}$ the marginal condition (2b) for the *momentum* entails consequences for the time variations of the individual values of the *position* variable, and these in their turn entail consequences for the statistical position distribution $P(\mathbf{q}) = \int P(\mathbf{q}, \mathbf{p})\,d\mathbf{p}$. Now for the chosen state vector the consequences on $P(\mathbf{q})$ of (14) and (2b) are not compatible with the stationarity of $P(\mathbf{q})$ required by the hypothesis of observability of $\mathbf{q}$ and by the marginal condition (2a) for the position. Indeed (14) and (2b) entail non-null z-components for the time variations of the (observable) $\mathbf{q}$

(15) $\qquad \dfrac{dq_z}{dt} = \dfrac{p_z}{K} = \pm|p_z| \neq 0.$

This entails that, if at some initial time t_0, (1.2a) is realized, throughout the future $t > t_0$ of this time the location with respect to oz of the maxima and minima of $P(\mathbf{q})$ keep reversing by a continuous process, with a time-periodicity

(16) $\qquad dt = \dfrac{K\,dq_z}{|p_z|} = \dfrac{Ki}{2|p_z|},$

where i is the distance at t_0 between two successive maxima of $P(\mathbf{q})$.

This example suffices for establishing Theorem 1. It shows that a joint probability $P(q, p)$ of observable beable values q, p which fulfils both marginal conditions (2) for any ψ, is a self-contradictory concept.

Since a joint probability $P(q, p)$ of beable q, p which fulfils both marginal conditions (2) is either unnecessarily restricted or self-contradictory and since, for a priori relevance to the reduction problem, the beable hypothesis for q, p has to be conserved, we conclude that *at least one of the two marginal conditions* (2) *has to be dropped.*

3.5. *Minimally Restricted Relevant Conditions of Consistency*

We admit by hypothesis that the object denominated one micro-system (S) does possess a continuously moving beable position and the corresponding beable momentum. Statistically this leads to the assumption, for any state vector ψ, of a corresponding joint probability of beable position and momentum variables. We shall now characterize this distribution so as to keep constantly faithful to the *minimality* of the model of a microsystem introduced by the mere assumption of a continuously moving beable position, while ensuring nevertheless a priori relevance to the reduction problem. Then, for the sake of minimality, we start out with a joint probability $P_\psi(q', p')$ where neither the position beable q' nor the momentum beable p' is asserted to be observable.

Condition for the momentum

We examine first the momentum distribution $\int P_\psi(q', p') \, dq' = P_\psi(p')$ because it seems less queer to admit that it is not observable, i.e. that in general it is different from the quantum mechanical momentum distribution: $(\int \mathscr{P}_\psi(q, p') \, dq = \mathscr{P}_\psi(p')) \neq |\Phi(p)|^2$ (Φ is the Fourier transform of ψ). For relevance to the reduction problem we have to admit that an individual act of momentum measurement relates the one preexisting beable momentum p' of the respective system, to the observed value p. This leaves (in general) an active role to the momentum measurement device $D(p)$, in agreement with Bohr's ideas: if λ is a parameter characterizing the state of $D(p)$, the observed value p is a function $P(p', \psi, \lambda)$ of p', ψ and λ, the form of this function (unknown) being fixed once ψ and a device $D(p)$ are given. Now, for any *physically realizable* ψ *and inasmuch as the quantum mechanical prediction* $|\Phi(p)|^2$ *is true for* ψ, we have to assume that statistically $p(p', \psi, \lambda)$ is obtained the number of times (normalized) $|\Phi(p)|^2$. This number can also be written $P_\psi(p')R_{D(p)}(\lambda)$ where $R_{D(p)}(\lambda)$ designates the statistical distribution of λ over the ensemble of the states of $D(p)$ realized for the individual acts of measurement which yield $|\Phi(p)|^2$, and where p', λ are taken the same as in the argument of $p(p', \psi, \lambda)$; indeed $P_\psi(p')$ and $R_{D(p)}(\lambda)$ are *independent* densities, since in every individual act of measurement interaction p' *preexists* to the interaction, by hypothesis. The necessity to label somehow the products $P_\psi(p')R_{D(p)}(\lambda)$ in relation with the observed values p, leads then to the mean condition

$$(17)_1 \qquad \iiint p(p', \psi, \lambda)P_\psi(q', p', t_0)R_{D(p)}(\lambda)\, dq'\, dp'\, d\lambda$$

$$= \iint p(p', \psi, \lambda)P_\psi(p', t_0)R_{D(p)}(\lambda)\, dp'\, d\lambda$$

$$= \int p|\Phi(p, t_0)|^2\, dp = \left\langle \psi(q, t_0) \left| \frac{\hbar}{i}\frac{\partial}{\partial q} \right| \psi(q, t_0) \right\rangle.$$

We have written explicitly the *constant* time t_0 elapsed since the state ψ has been prepared for each individual S, when the corresponding individual act of measurement interaction between $D(p)$ and S begins: thereby we emphasize that the numerical equality $(17)_1$ does not depend on the time evolution of the measurement interactions, neither on their functional form nor on their duration; it depends exclusively on the connection between their *result* (second member) and circumstances which *precede* them (first member, t_0).

But, beyond the numerical aspects, it is important to understand clearly the conceptual content of the integrand from the first member of $(17)_1$: while the values of the functional $p(p', \psi, \lambda)$ are the observed values p from $\int p|\phi(p, t_0)|^2\, dp$, the functional form of $p(p', \psi, \lambda)$ represents the hypothetical – individual and deterministic – process which leads from one beable value p' possessed by the supposed momentum beable of the system, at the time t_0 when the measurement interaction began, to the observed value p, defined at another time, by a coordinate attached to a macroscopic part or aspect ('pointer') of $D(p)$. The presence of the parameter λ in the argument of $p(p', \psi, \lambda)$ stresses the assumption that this individual process depends – besides p' *and*

ψ – also on the state of $D(p)$, throughout the time interval taken by the measurement interaction. The definition of this state of $D(p)$ introduces a macroscopic potential (constant or null, for p) which is different in general from the macroscopic potentials having commanded the Schrödinger evolution of ψ from the moment from which ψ has been prepared until t_0 when the act of measurement began. Thus the exact meaning hypothetically assigned to $p(p', \psi, \lambda)$ is this: it represents one *individual* member of a virtual statistical ensemble of p-measurement evolutions, *globally* corresponding to the Schrödinger evolution of the state vector of the 'system + $D(p)$', during the p-measurement interaction. We finally note that, for the sake of maximal generality, we conceive that the functional form of $p(p', \psi, \lambda)$ might depend upon the particular $D(p)$ device utilized. Two different devices $D_1(p)$ and $D_2(p)$ can be conceived to introduce in general two different functional forms $p^{(1)}(p', \psi, \lambda)$ and $p^{(2)}(p', \psi, \lambda)$ and two different distributions $R_{D_1(p)}(\lambda_1)$ and $R_{D_2(p)}(\lambda_2)$. But then a certain correspondence has to be also assumed between $p^{(1)}$, $R_{D_1(p)}$ and between $p^{(2)}$, $R_{D_2(p)}$, such that statistically, in a given $\psi(q, t_0)$ prepared for each microsystem S, both $D_1(p)$ and $D_2(p)$ shall create any given observed value p, with the same relative frequency $|\Phi(p, t_0)|^2$.

Condition for the position
We require for the position the same type of consistency condition as for the momentum, in order to conserve the minimality of the demanded restrictions:

$$(17)_2 \qquad \iiint q(q', \psi, \lambda) P_\psi(q', p', t_0) R_{D(q)}(\lambda) \, dq' \, dp' \, d\lambda$$

$$= \iint q(q', \psi, \lambda) P_\psi(q', t_0) R_{D(q)}(\lambda) \, dp' \, d\lambda$$

$$= \langle \psi(q, t_0) | q \psi(q, t_0) \rangle = \int |\psi(q, t_0)|^2 q \, dq$$

(obvious notations). All the comments concerning $(17)_1$ are transposable for $(17)_2$. We make now an important remark concerning $(17)_2$:

In the first place, this *mean* condition for the position, in contradistinction to the marginal condition (2a), leaves open the possibility that the beable position q' of one microsystem $S^{(\psi)}$, lies outside the support of ψ. However shocking it might seem, this possibility cannot be excluded since the purely predictive formalism of quantum mechanics introduces no assertion whatever concerning the way in which the only observationally described object $S^{(\psi)}$ 'exists' independently of observation. However (2a) subsists inside $(17)_2$ as a particular possibility. In consequence of Theorem 1, one at least of the two marginal conditions (2a) and (2b) has to be dropped, but not necessarily both. Since we have dropped the marginal condition for the momentum, we remain free for the moment to assume that the marginal condition for the position is always true. But it will appear that this apparently so natural assumption has a heavy price, if all the quantum mechanical predictions are true.

The other mean conditions (macroscopic dynamical. quantities, quantum mechanical dynamical operators, beable dynamical quantities)
For dynamical quantities more complex than q and p, most of the mean conditions posed so far in connection with joint probability attempts – and then criticized – have a structure which does not resist a closer analysis. Given a *macroscopic* classical dynamical quantity $f_m(q, p)$, the corresponding *beable* dynamical quantity of a microsystem $S^{(\psi)}$ is usually conceived in a way which violates the *minimality* of the model of a microsystem introduced by the mere hypothesis of a continuously moving beable position: the beable which corresponds to $f_m(q, p)$ is brutally identified with $f_m(q, p)$ and thereby the naïve atomistic model, made obsolete by de Broglie more than fifty years ago, is implicitly reintroduced. Moreover, the fact that a measurement interaction in general *modifies* the beable characteristics of a microsystem, there*from* yielding an observed value, is not taken into account. Such unanalyzed steps lead to mean condition of the type

$$\iint f_m(q, p) P_\psi(q, p) \, dq \, dp = \left\langle \psi | f_{m, QM}\left(q, \frac{\hbar}{i} \frac{\partial}{\partial q}\right) \psi \right\rangle,$$

($f_{m, QM}$ is the quantum mechanical operator for f_m), and then these are found unsatisfactory, which indeed they are. Before going over to locality analyses we shall express these criticisms more detailedly. This will enable us to specify what mean conditions, for any quantity, can be imposed upon a joint probability both minimally restricted and relevant.

We begin by recalling a well-known fact concerning the time evolution conceivable for a joint probability of beable q', p'. Since

$(17)_1$ and $(17)_2$ are required for $P_\psi(q', p')$ at any time, P_ψ has to perform a time evolution compatible with the Schrödinger evolution of the corresponding ψ. This evolution admits a newtonian representation in consequence of the kinematic definition $p' = K(dq'/dt)$ assumed for each p'. Indeed – by definition – the time evolution of P_ψ is newtonian if it is describable by an equation of the form

$$(18) \qquad \frac{\partial P_\psi(q', p')}{\partial t} = \frac{p'}{K} \frac{\partial P_\psi(q', p')}{\partial q} + F \frac{\partial P_\psi(q', p')}{\partial p},$$

where the symbols q' and p' are pairwise connected precisely by the kinematic relation $p' = K(dq'/dt)$, while the time variation of p' is equated, by application of the fundamental newtonian postulate, to a convenient 'total force' F, classical or *not*,

$$(19) \qquad \frac{dp'}{dt} = F,$$

(this force can be conservative, or dissipative, or a sum of a conservative term and of a dissipative term; only in the first case it is derivable from a potential function, and then (18) acquires a hamiltonian form). Now, it is well established that, given the Schrödinger evolution of ψ determined by some macroscopic potential $V_m(q)$, it is in general *not* possible to find a newtonian evolution (18) for an attempted joint probability P_ψ, if F in (19) is required a priori identical with the macroscopic force $F_m = -\text{grad } V_m(q)$: proofs of this impossibility are contained implicitly, but rather obviously, in the text-book studies of the WKB approximation as well as in Feynman's path integral approach[3] or in de Broglie's and Bohm's hidden variable attempts. Thus F in (18) has to be conceived as an unknown non-macroscopic force which cannot be posed, but which has to be determined consistently with the Schrödinger evolution of ψ, as a functional of $V_m(q)$ *via* $\psi(V_m(q))$. This functional would probably yield the most specific descriptive element of a non-naïve model of a microsystem.[1] If, on the contrary, F in (18) is decreed to be identical to $F_m = -\text{grad } V_m(q)$, any hope for a joint probability $P_\psi(q', p')$ performing a time evolution consistent with ψ – for any ψ – is thereby banished.

On the basis of this remark it will now be easy to understand that

PROPOSITION 2.   Given a *macroscopic* classical dynamical quantity $f_m(q, p)$, a corresponding *beable* classical dynamical quantity does not necessarily exist; if it does exist, then it is in general different from the corresponding $f_m(q, p)$, so that it cannot be found by reversing the correspondence rule which led from $f_m(q, p)$ to the respective quantum mechanical operator $f_{QM}(q, (\hbar/i)(\partial/\partial q))$.

Proof.   Again we produce an example. Consider the macroscopic dynamical quantity total energy $f_m(q, p) = H_m(q, p) = p^2/2m + V_m(q)$. Consider also *one* individual microsystem $S^{(\psi)}$. What can be said concerning a beable total energy of $S^{(\psi)}$? With our previous assumptions $S^{(\psi)}$ possesses a beable position and a corresponding beable momentum $p' = K(dq'/dt)$. One can then form for $S^{(\psi)}$ a kinetic energy p'^2/K (where K is not identical to the mass m of $S^{(\psi)}$, a priori). But in order to preserve for a joint probability $P_\psi(q', p')$ attempted for $S^{(\psi)}$, the possibility of a time evolution compatible with that of ψ, the force $F = dp'/dt$ which – by newtonian postulate – is equated to dp'/dt, has to be in general different from the macroscopic force $F_m(q) = -\text{grad } V_m(q)$, $F'(q') \neq F(q)$. If moreover $F'(q')$ is not conservative, then $S^{(\psi)}$ simply does not possess a beable hamiltonian, notwithstanding the fact that the time evolution of ψ is expressed by a hamiltonian (operational) formalism [4]. If on the contrary $F'(q')$ also does derive from a potential, this potential $V'(q') \neq V_m(q)$ is in general different from $V_m(q)$; then $S^{(\psi)}$ does possess a beable hamiltonian $H_b = p'^2/2K + V'(q')$ but this is different from the macroscopic hamiltonian $H_m = p^2/2m + V(q)$ to which corresponds the hamiltonian evolution operator for ψ: $H_{QM} = -(\hbar/2m)(\partial^2/\partial q^2) + V(q)$. Replacement in H_{QM} of $(\hbar/i)(\partial/\partial q)$ by p, and of the multiplicative operator $V(q)$ by the function $V(q)$, yields back H_m but not $H_b (\neq H_m)$.

This example suffices for showing that mean conditions of the form

$$\iint f_m(q, p) P_\psi(q, p) \, dq \, dp = \langle \psi | f_{m,QM}(q, (\hbar/i)(\partial/\partial q)) \psi \rangle.$$

are not significant. (In particular such a mean condition for the potential energy itself

$$\iint V_m(q) P_\psi(q, p) \, dq \, dp = \langle \psi | V_m(q) \psi \rangle,$$

is the very *definition* of a naïve, atomistic postulate on the structure
of the microreality). P_ψ and ψ cannot be purely algorismically treated
as if they were both fit for relevantly calculating means of *any* and
same functions. P_ψ can yield relevant means for beable values only
while ψ is relevant for calculating means of observed values only.
Park and Margenau have explicitly contested – on logical grounds –
the relevance of mean conditions written with the macroscopic functions $f_m(q, p)$[5]; Proposition 2 gives a more physical reason of this
irrelevance. But obviously there exists a much more radical objection:
given a quantum mechanical operator $f_{m,QM}$ corresponding to the
macroscopic dynamical quantity f_m, even if the respective beable
quantity both does exist and is distinguished from f_m, not *its* mean value
is relevant to the reduction problem, but the mean engendered by it via
the measurement interactions, which depend also on the measurement
device. Bohr's views on measurement were very profound, each act of
measurement modifies preexisting characteristics of the system, bringing out from it observed values of *other*, only operationally defined
'quantities of the system'.

Then all that *can* be required of a joint probability $P_\psi(q', p')$ of beable
q', p' is to have an analytic expression such as to be *compatible* with
mean conditions of the type $(17)_1$ and $(17)_2$, for *any* quantum mechanical
dynamical observable w, at any time, i.e.

$$(17) \qquad \iiint w(q, p, \psi, \lambda) P_\psi(q', p', t_0) R_{D(w)}(\lambda) \, dq' \, dp' \, d\lambda$$
$$= \left\langle \psi(q, t_0) | f_{QM,w}\left(q, \frac{\hbar}{i}\frac{\partial}{\partial q}\right) \psi(q, t_0) \right\rangle = \int w |C^\psi(w, t_0)|^2 \, dw,$$

where all the notations have obvious meanings by analogy with $(17)_1$. All
the comments concerning $(17)_1$ can be transposed to (17), which includes
now $(17)_1$ and $(17)_2$. We can rewrite (17) in a form more specifically
connected with the dynamical observable w: Given one $S^{(\psi)}$ we denote
globally by a *unique* parameter w' all the beable characteristics of $S^{(\psi)}$
which contribute, with ψ and λ, to the creation of the observed value w
when one act of w-measurement is performed on $S^{(\psi)}$. These characteristics can be conceived as defined *at q'* since q' designates the beable
element of $S^{(\psi)}$ to which a beable dynamics is assigned. Then statistically
the joint distribution $P_\psi(q', p', t_0)$ defines a corresponding joint distribution $\Pi_\psi(w', q', t_0)$. Rewriting of $w(q', p', \psi, \lambda)$ in function of w'

yields a function of a new functional form $w(w', \psi, \lambda)$ but the values of which continue to be the observed values w, and for which all the considerations made for the particular case of $p(p', \psi, \lambda)$ from $(17)_1$ are valid. So (17) becomes

$$(17)' \quad \iiint w(w', \psi, \lambda) \Pi_\psi(q', w', t_0) R_{D(w)})(\lambda) \, dq' \, dw' \, d\lambda$$

$$= \langle \psi(q, t_0) | f_{QM,w} \psi(q, t_0) \rangle = \int w |C^\psi(w, t_0)|^2 \, dw.$$

The critical remarks which led to condition (17)' show that all the theorems of impossibility (like that of von Neumann concerning simultaneous measurements of quantities with non-commuting quantum mechanical operators ([16], pp. 255–230), or that of Kochen and Specker[7], as well as all the investigations on joint probabilities based on correspondence rules with the quantum mechanical operators (Moyal[8], Bass[9], Cohen[10])) must be carefully reconsidered. Indeed: If the quantum mechanical operators of two quantum mechanical dynamical observables w_1 and w_2 do not commute, this expresses – by definition – the fact that the quantum mechanical measurement processes yielding the quantum mechanical operational definitions for w_1 and w_2, *cannot* be realized simultaneously in one individual act of measurement. Hence, when one examines the question of the "simultaneous measurability of two observables w_1 and w_2 associated with two non-commuting quantum mechanical operators", *ipso facto* a *non-quantum-mechanical* operational definition is now envisaged for at least one of these two quantities, namely a definition such that, now, the two measurement processes conceived *shall* 'commute' (shall *be* simultaneously realizable in one individual act of measurement). In other terms, this problem cannot concern the *same* initial pair of observables w_1, w_2; it can only concern *another* pair, where at least one member is changed. This does not at all mean that the problem is absurd. Nothing hinders the conception that one given beable property w' assigned to a system can be connected with observable facts via several different operational definitions. But there is no reason then to expect for such different operational definitions the same statistical distribution of observed results; different observed statistical distributions have to be expected for them, in general. All these observable distributions are equally acceptable for 'describing' the unique intrinsic distribution of values supposed for the beable quantity w' assigned to the studied system, under the sole condition that each one of the observable distributions be related in some definite – even though specific – way with this unique intrinsic distribution. These considerations entail that when the question of simultaneous measurability is examined, one at least of the two $w(w', \psi, \lambda)$ functionals intervening, describes an individual measurement evolution that is somehow *not compatible* with the quantum mechanical operator for w. There is then no reason whatever to require the equality (17)' when such a $w(w', \psi, \lambda)$ acts (as Park and Margenau[5] did, as well as von Neumann[6]). Furthermore, there is no reason whatever either for subjecting the functional forms $w(w', \psi, \lambda)$ from (17)' to structural correspondence rules with the quantum mechanical operators associated to the w-quantities, nor for requiring for these functionals an algebra identical to that of the quantum mechanical operators. The $w(w', \psi, \lambda)$ from (17)' represent *processes*, and these, moreover, are posed to be *individual*: this is the essential feature of any attempt of a 'causal' solution to the reduction problem. Whereas any quantum mechanical w-operator is defined in direct formal connection with the function $f_{m,w}$ describing the classical macroscopic w-*quantity*; this operator, moreover, is in a one-to-one relation with a whole family of eigenvectors ϕ_{jw}, to each one of which a joint probability attempt assigns already a *statistical* significance, as it can be seen for instance by writing (17)' for a ϕ_{jw} and by comparing the contents of the two members:

$$\iiint w(w', \psi, \lambda) \Pi \phi_{jw}(q', w', t_0) R_{D(w)}(\lambda)\, dq'\, dw'\, d\lambda$$
$$= \langle \phi_{jw}(w, t_0) \,|\, f_{QM,w} \phi_{jw}(q, t_0) \rangle = w_j$$

(w_j is the eigenvalue corresponding to ϕ_{jw} of the quantum mechanical operator $f_{QM,w}$). It simply is not physics to impose upon the $w(w', \psi, \lambda)$ a priori formal constraints. The relevant constraints have to be deduced by means of very analyzed physical criteria brought forth by an improved insight in the joint probability problem. We believe that such an insight cannot be obtained as long as only surface probabilistic relations, connecting probability *measures* alone, are stated explicitly, while the corresponding relations between the *events* concerned by these measures are left more or less in the dark. All the various probability *spaces* which intervene – quantum mechanical probability spaces and joint probability spaces – have to be studied in their entirety and with their interplay at all the levels ('conditions' defining the 'experiment', elementary events brought forth, field on these, measure on the field), in order to acquire a precise and complete perception of the deep structure of the joint probability problem [12].

3.6. Generalization to Any Relevant Hidden Distribution

The 'dynamical' observables associated to S correspond – by their operators $f_{QM}(q, \hbar/i(\partial/\partial q))$ – to the classical dynamical quantities, which are all defined as functions $f(q, p)$ of the position and the momentum. Therefore the concept of a joint probability $P_\psi(q', p')$ of a beable position and a beable momentum variable seems a 'natural' concept for expressing the consistency condition (17), to be required for the quantum mechanical 'dynamical' observables associated to S. This joint probability concept, however, cannot yield a direct representation of the 'field-like' beable properties tentatively conceivable for a microsystem; it reflects such properties only indirectly, via the non-classical forces necessary (in general) in the time-evolution law (18), if one wants to preserve the possibility of some compatibility with the Schrödinger evolution of ψ (pp. 125–128). Therefore the joint probability concept $P_\psi(q', p')$ is not appropriate for expressing a consistency condition concerning the quantum mechanical observables of S to which no classical function $f(q, p)$ corresponds (charge, spin component on a given direction). Indeed, for such an observable it would be a priori restrictive to pose that the beable properties w' of S which lead to the observed values w (via the process $w(w', \psi, \lambda)$) are defined at q', as it has been assumed for the dynamical quantities considered in (17). Therefore we generalize (17) and (17)' by making use of a hidden distribution $P_\psi(\mu')$ instead of the joint probability $P_\psi(q', p')$, and of a functional $w(\mu', \psi, \lambda)$ instead of $w(q', p', \psi, \lambda)$, μ' being a generalized hidden variable which designates globally any sort of beable properties assigned to S and conceived to lead to the observed value w via the interaction process described by $w(\mu', \psi, \lambda)$:

$$(17)'' \quad \iint w(\mu', \psi, \lambda) P_\psi(\mu', t_0) R_{D(w)}(\lambda) \, d\mu' \, d\lambda =$$

$$= \langle \psi(q, t_0) | w_{QM} \psi(q, t_0) \rangle = \int w c^\psi(w, t_0) \, dw$$

(w_{QM} in the second member is the quantum mechanical operator of the observable w, connected or not with a classical function $f(q, p)$). Thus (17)″ englobes now (17) and (17)′: we have finally obtained a condition applying to any hidden distribution – a joint probability, or some other distribution – which is both minimally restricted and still relevant to the reduction problem.

3.7. *Methodological Attitude Concerning the Consistency Condition* (17)″

We want to stress a methodological attitude to which we attach a fundamental importance: we assign to the condition (17)″ a symmetric role with respect to quantum mechanics and with respect to a hidden variable attempt, we do not subordinate inconditionally the hidden variable attempts to quantum mechanics.

The conditions of consistency attempted so far have all presupposed the exceptionless validity of the quantum mechanical predictions, at least in the domain of atomic dimensions and newtonian energies. However the fact that a hamiltonian operator can be written does not ensure the physical realizability of its potential term, neither that, a fortiori, of the corresponding Schrödinger time evolution-law. If now a physically realizable potential and the corresponding evolution-law are considered, the mathematically possible ψ-solutions do not all correspond to physically realizable boundary conditions. And if a physically realizable ψ is considered, very paradoxically, the quantum mechanical 'observables' of the system do not all possess a unanimously admitted and physically realizable operational definition, so that the corresponding prediction is not always verifiable (the most striking example of this sort concerns the fundamental 'observable' momentum: in a state ψ which is not an eigenstate of the momentum, according to the orthodox theory of measurement a rigorous measurement of the momentum for $\psi(t)$ yields the observed results at t' such that $(t' - t) \sim \infty$ (time of flight method)). Finally if one considers a physically realizable ψ and an observable for which an admitted operational definition does exist and the results of its application are observable, then the corresponding quantum mechanical prediction might never have been verified.[2] But a priori restrictions corresponding to unrealizable, or to non-verifiable, or to non-verified features of the quantum mechanical description, are likely to introduce fatal malformations into a joint probability attempt. For these reasons, while requiring the conditions (17)', we have no rigid preconception. Even these minimally restricted conditions of consistency are demanded only for physically realizable state vectors and we shall keep in mind the two important problems of the verifiability and of the verification of the involved quantum mechanical predictions. In this way, while quantum mechanics imposes restrictions upon the acceptable joint probability, this, in its turn, can play the role of a test concept concerning the quantum mechanical description. This attitude is novel and it is characteristic of our approach.[3,4]

3.8. One-System Non-Locality

Joint probability framework
We place ourselves inside the joint probability framework, which afterwards we shall leave. The joint probability defined by the minimal condition (17)' might seem a very weak concept, unable to lead to any definite conclusion for some problem. But we shall now show that in fact this minimally restricted, while still relevant, concept of a joint probability is strong enough for entailing a problem of physical non-locality inside the one-system formalism of quantum mechanics.

Preliminaries. We make first two remarks:
(1) According to quantum mechanics, if a microsystem S is at some time t in a superposition state $\psi_{ab} = a\psi_1 + b\psi_2$, whose support I in the physical space is a non-connected union $I = I_1 \cup I_2$ of two spatially disjoint intervals $I_1, I_2 (I_1 \cap I_2 = \emptyset)$, then it is possible to prepare the state ψ_1 for S, out of the state ψ_{ab}, namely by suppressing at t on I_2 – with the help of an obturator or filter acting on I_2 – the characteristics of S described by the term $b\psi_2$ of ψ_{ab}. Indeed, if $\Delta t_{pr} = (t_{pr} - t)$ is the time taken by the action of the filter or obturator ('preparation' time), from $t_{pr} = t + \Delta t_{pr}$ on, the state vector to be assigned to S is ψ_1 alone, renormalized to unity. This type of preparation is particularly interesting from our viewpoint because it asserts a relation between a physical – but not *observational* – operation, carried out with the help of a macroscopic device at the location (namely I_2) of a *descriptive* element (namely $b\psi_2$), and a certain physical modification of the 'state' assigned to the object designated by S, possibly entailing changes in the *beable* properties assumed for this object. Even though the quantum theory asserts nothing whatsoever concerning the location in the physical

space, outside the periods of observation, of the objects described by this theory, the possibility of a preparation of the type specified above might contain some implications as to where this object can 'exist' outside the periods of observation, according to a joint probability theory fulfilling conditions of consistency with quantum mechanics.

(2) As we have already pointed out, the quantum mechanical momentum observable has a peculiar operational definition, namely the time-of-flight method. According to this definition the measurement begins at a moment t_0 by the suppression of all external fields, if they existed, while the interaction with a material *registering* device $D(p)$ (which yields, directly, a *position* value) is relevant only if it occurs at another time t, such that $t - t_0 = \Delta t(p) \sim \infty$. The complete measurement interaction consists here of the passage of the infinite period $\Delta t(p)$ + the final registering interaction with $D(p)$. Now, the infinite value thus required for $\Delta t(p)$ introduces ambiguities at the level of a joint probability theory: in the first place, it rules out a *rigorous* verifiability of the quantum mechanical prediction for the momentum spectra. Moreover, not even an approximate verification of this prediction seems ever to have been made effectively for the various types of preparable states ψ (in particular for the superposition states $\psi_{ab} = a\psi_1 + b\psi_2$ with non-connected support, or with connected support (interference)). Therefore, faithful to the agnostic attitude we choose, we reserve our opinion as to the circumstances in which the consistency condition $(17)_1$ concerning the momentum has to be required. In the second place, in the case of a free Schrödinger evolution of ψ, the quantum mechanical operational definition of the momentum observable permits a *degenerate* relation between the observable p-spectrum asserted by quantum mechanics and the instantaneous structure of the hypothetic beable distribution of a hidden momentum $P_\psi(p') = \int P_\psi(q', p') \, dq'$, corresponding to the joint probability measure from $(17)_1$. Indeed the quantum mechanical p-spectrum is an invariant of a free Schrödinger evolution. Then the whole family of different instantaneous structures taken on by $P_\psi(q', p', t_0)$ from the left member of $(17)_1$ when time translations change the t_0 considered, correspond to one same quantum mechanical p-spectrum in the right member of $(17)_1$, if ψ has a free Schrödinger evolution. However, as soon as the beable properties assigned to the object S are different from those of a material point (which seems rather unavoidable, as the remarks on pp. 125–128 show), the beable momentum distribution $P_\psi(p') = \int P_\psi(q', p' = K \, dq'/dt) \, dq'$

can – in general – *change* during a free evolution of ψ, in consequence of the kinematic definition $p' = K(dq'/dt)$ of the beable momentum. Once more an illustration is yielded by the superposition states, namely those which, like $\psi(q, t) = (1/\sqrt{2})\phi_{p_1}(q) + (1/\sqrt{2})\phi_{p_1}(q)$ from the proof of Theorem 1 (pp. 120–122) take on successively, during their Schrödinger evolution, a connected support first, and then a non-connected support (or vice versa) [13]. The preceding remarks apply as well to any function of the momentum alone. But consider now quantities w not depending on the momentum alone (kinetic momentum, projections of the kinetic momentum, total energy). The quantum mechanical operational definitions of such quantities consist of procedures where the time at which the interaction itself between one $S^{(\psi)}$ and a material device $D(w)$ begins, coincides with the time t_0 from (17)' at which what is called 'measurement' as a whole begins. Moreover, the duration $\Delta t(w)$ required in principle for such a measurement *is not infinite*. The preceding remarks concerning the quantities depending on the momentum alone do not apply to these other quantities. We shall now show that:

THEOREM 2.   If it is assumed that the beable properties assigned at a time t to the object denominated one system $S^{(\psi)}$ cannot lie outside the support in the physical space of the quantum mechanical state vector $\psi(t)$ associated to $S^{(\psi)}$, then even the minimally restricted joint probability concept from (17)' is unable to ensure a local deterministic solution to the reduction problem, for any state vector and any dynamical observable.

This theorem will be proved by giving an example. Our choices for an example are the following ones:

For the reasons given in the preliminary remarks (b) we consider a quantity w of which the quantum mechanical operational definition involves a *finite* measurement interaction time

(20) $\Delta t(w) < \infty$.

Furthermore, at some initial time t_i, we consider the three state vectors $\psi_1, \psi_2, \psi_{ab} = a\psi_1 + b\psi_2$ such that the supports in the physical space, I_1 and I_2, of – respectively – ψ_1 and ψ_2, are disjoint. The distance d_{12} separating the two nearest points of I_1 and I_2 is subject to a condition, namely: we denote by Δt_{pr} the time-interval necessary for preparing for S the state described by ψ_1 out of the state described by ψ_{ab}, by the method mentioned in remark (a) (i.e., Δt_{pr} is the time-interval, *finite*, taken by an obturator or a filter for suppressing on I_2 the characteristics of S described in ψ_{ab} by the term $b\psi_2$). The moments $t_i < t_0 < t$ are chosen such that $\Delta t_{pr} = t_0 - t_i$, $\Delta t(w) = t - t_0$. We denote $\Delta t_{pr} + \Delta t(w) = \Delta t$ and we require

(21) $d_{12} > c\Delta t,$

where c designates the velocity of light.

With these choices we can now develop the proof of Theorem 2. We shall first show that

LEMMA.   The product $w\Pi_{ab}$ intervening in the integrand from the first member of the condition (17)′ written for ψ_{ab}, has a *mathematically non-local* dependence on ψ_{ab}.

Proof. The condition (17)′ written for ψ_{ab}, ψ_1, ψ_2, yields (with obvious notations)

(22) $$\iiint w(w', \psi_{ab}, \lambda)\Pi_{ab}(q', w', t_0)R_{D(w)}(\lambda)\, dq'\, dw'\, d\lambda$$

$$= \langle \psi_{ab}(q, t_0)|f_{QM,w}\psi_{ab}(q, t_0) = |a|^2 \int w|C^{(1)}(w)|^2\, dw$$

$$+ |b|^2 \int w|C^{(2)}(w)|^2\, dw + a^*b \int w(C^{(1)}(w))^*C^{(2)}(w)\, dw$$

$$+ ab^* \int w(C^{(2)}(w))^*C^{(1)}(w)\, dw,$$

(23) $$\iiint w(w', \psi_1, \lambda)\Pi_1(q', w', t_0)R_{D(w)}(\lambda)\, dq'\, dw'\, d\lambda$$

$$= \langle \psi_1(q, t_0)|f_{QM,w}\psi_1(q, t_0)\rangle = \int wC^{(1)}(w)|^2\, dw,$$

(24) $$\iiint w(w', \psi_2, \lambda)\Pi_2(q', w', t_0)R_{D(w)}(\lambda)\, dq'\, dw'\, d\lambda$$

$$= \langle \psi_2(q, t_0)|f_{QM,w}\psi_2(q, t_0)\rangle = \int w|C^{(2)}(w)|^2\, dw,$$

Let us admit tentatively the hypothesis conditionally contained in the formulation of Theorem 2, namely

(h) the beable properties assigned at a time t to what is named one system $S^{(\psi)}$ cannot lie outside the support of $\psi(q, t)$ in the physical space.

Consider now the product $w\Pi_{ab}$ from the left member of (22). The hypothesis (h) entails that this product is null outside the support $I = I_1 \cup I_2$ of $\psi_{ab}(q, t_0)$, because the probability measure $\Pi_{ab}(q', w', t_0)$ is null for $q' \notin I$. Then the non-connected structure chosen for $I = I_1 \cup I_2$ (namely $I_1 \cap I_2 = \emptyset$) entails that the product $w\Pi_{ab}$ from (22) is a sum of two terms

(25) $\quad w(w', \psi_{ab}, \lambda)\Pi_{ab}(q', w', t_0) = w(w', \psi_{ab}, \lambda)\Pi_{ab,I_1}(q', w', t_0)$
$\qquad\qquad + w(w', \psi_{ab}, \lambda)\Pi_{ab,I_2}(q', w', t_0),$

(obvious notations) of which one is null for any given q', since *one* $S^{(\psi_{ab})}$ possesses *one* beable 'position' property, so that either $q' \in I_1$ and then $q' \notin I_2$, or vice versa. However, confrontation of (25) with (22), (23), (24) shows that in general

(26)
$\quad w(w', \psi_{ab}, \lambda)\Pi_{ab,I_1}(q', w', t_0) \neq w(w', \psi_1, \lambda)\Pi_1(q', w', t_0),$
$\quad w(w', \psi_{ab}, \lambda)\Pi_{ab,I_2}(q', w', t_0) \neq w(w', \psi_2, \lambda)\Pi_2(q', w', t_0),$

because the sum of the two last 'interference' terms in the second member of (22) is not null for any ψ_1, ψ_2, a, b, and w. It *is* null for the particular case $w = g(q)$ (because of $I_1 \cap I_2 = \emptyset$) so that for these quantities the non-equalities (26) transform into equalities. But for $w \neq g(q)$ the term $w\Pi_{ab,I_1}$ from (25) depends – as the first non-equality (26) shows – on the whole superposition state vector $\psi_{ab} = a\psi_1 + b\psi_2$, even though this term is defined on I_1 alone and even though $I_1 \cap I_2 = \emptyset$, the distance d_{12} which separates I_1 from I_2 being moreover arbitrarily big, as (21) permits. The symmetric argument holds for the term $w\Pi_{ab,I_2}$ from (25). In *this* sense the product $w(\psi_{ab})\Pi_{ab}$ from (22) has a mathematically non-local dependence on ψ_{ab}, q.e.d. The lemma proved above generalizes to any relevant joint probability from (17)' the mathematical non-locality of Wigner's joint probability (1) (expressed by (11), (12), (13)).

But $w(w', \psi, \lambda)$ designates a *process*, which, in addition, takes a non-null time-interval $\Delta t(w)$. Therefore, in order to investigate whether or not the mathematical non-locality brought into evidence above does involve *physically* non-local phenomena, time has to be taken into account also. This is what one shall do now:

The hypothesis (h) has a rather obvious consequence, namely:

(c) Given one system S to which a quantum mechanical state vector $\psi(t)$ is associated at the time t, throughout any *local* process which involves the system S during a period $\Delta t = t' - t$, the transforms by this process of the beable properties assigned to S at the initial time t, remain confined inside the portion corresponding to Δt of the light-cone of the support I of $\psi(t)$. (This formulation holds with respect to any given space-time referential and, whether or no, from t *on* the quantum theory continues to associate an individualized state vector with S.)

Consider then a statistical ensemble of systems S for each one of which, at the constant time t_i after the preparation of ψ_{ab}, the new state ψ_1 is prepared out of ψ_{ab}, and then a w-measurement is performed on $S(\psi_1)$ (the choices (20), (21) being fulfilled). Under these conditions, if the consistency relation (22) for ψ_{ab} is satisfied, then the condition (23) for ψ_1 is violated, unless some non-local effects take place. Indeed:

The consequence (c) of (h) together with (20) and (21) entail that throughout the time-interval $\Delta t = \Delta t_{pr} + \Delta t(w)$ taken by the global process [preparation for S of the state described by ψ_1 out of the state described by $\psi_{ab} + w$-measurement on $S^{(\psi_i)}$] the transforms by this process of the beable properties assigned to S at t_i remain confined inside two disjoint and space-like separated space-time domains. But according to (26) the consistency condition (23) for ψ_1 can be fulfilled only if the product $w(\psi_{ab})\Pi_{ab,I_1}$ changes into the different product $w(\psi_1)\Pi_1$. This is a required *statistical* change, but it can come about only if *individually* the beable properties realized for each $S^{(\psi_{ab})}$ on I_1 at t_i, undergo during Δt a transformation different – in general – from the transformation that would have taken place if the state of that S would have continued throughout Δt to be described by ψ_{ab} (i.e. in absence of the action on I_2 of an obturator or filter). In other words, each one action of preparation of a state ψ_1 out of a state ψ_{ab} for one S, even though it takes place on I_2, must – in general – somehow cause a *change*, and *during* Δt, of the individual properties of that S on I_1. Now, if such a change does indeed happen, it can be only non-local, since the portions corresponding to Δt of the light cones of I_1 and of I_2 are two space-like separated space-time domains. While, if the specified change does not happen, (23) for ψ_1 cannot be fulfilled, in consequence of the first non-equality (26). This example suffices for proving Theorem 2.

Quite independently of any experimental investigation which it might suggest, this conclusion is a theoretical fact.

General hidden variables framework. Theorem 2 can be generalized for any hidden distribution fulfilling the minimal condition (17)″, and for any observable, as we have shown elsewhere [14]. Thus, when grasped synthetically, the conceptual situation is this: No hidden variable distribution can ensure a *local* deterministic solution to the reduction problem, *if* the object denominated 'one microsystem' cannot 'exist' outside the support in the physical space of the quantum mechanical state vector associated to it. Thus we have been led to a direct confrontation between the one-system quantum mechanics, causality, relativity, and the question *where* the object named 'one microsystem' does 'exist'. The concept of hidden variables has played the role of a revelator of this confrontation. This shows the methodological force of the hidden variables concept.

3.9. *Comparison with Two-System Non-Locality*

In the one-system locality theorem proved above, the question of the relation between the beable location of '*S*' in the physical space and the support of the quantum mechanical state vector of '*S*', plays an essential role; while J. S. Bell, who discovered the locality problem [2], has brought forth, with the help of his well-known two-system example, a pure and striking confrontation between quantum mechanical predictions, causality and relativity, where no explicit use is made of the question mentioned. In this connection we want to make two remarks.

In the first place: when '*S*' designates 'one system' only one mark on a measurement device can be registered for each '*S*'. This is what necessitates the explicit introduction of the hypothesis (h) in the demonstrations on one-system non-locality. However, if not a demonstration, a one-system alternative for experimental investigation on locality can be formulated without use of (h). Indeed, one can obtain a conclusion by exclusively taking into account the space-time coordinates of macroscopic events, namely the action of an obturator and the registration of a mark on a measurement device: even though quantum mechanics does not *predict* where and when a mark will be registered, a posteriori this mark is always found with *some* definite space-time coordinates. If, for each individual registration, these coordinates are found to be separated space-like from those of the action of the obturator (following the conditions of the proof of Theorem 2) and if, nevertheless, statistically, the quantum mechanical distribution for ψ_1 is found when the obturator is used, while, when not, the distribution predicted for ψ_{ab} is found, then there is non-locality.

In the second place: when it is tried to define the significance to be assigned to the various possible results of the experiments for verifying Bell's inequality, the question of where the object named a 'two system' does beably 'exist' comes into play irrepressibly, raising novel and fundamental problems [15], even though it is absent from Bell's demonstration, at least explicitly. (Implicitly it must somehow intervene, since the location of the two registering devices used is not chosen *independently* of the maxima of the presence probability for the two 'parts' of 'S', calculated with the help of the state vector of 'S'.)

From these remarks we conclude that the question of the relation between the beable location of 'S' in the physical space and the support of the quantum mechanical state vector of 'S' plays in fact an essential role in any locality problem, no matter whether 'S' designates 'one system' or 'two systems' and notwithstanding the formal descriptive differences.

In this perspective, the explicit presence of this question in the one-system demonstration appears as a specific and interesting feature, drawing particular attention to the relations between reality and the descriptive language of quantum mechanics.

3.10. *Experimental Study*

Theorem 2 and its generalization suggest an experimental study which we shall now indicate.

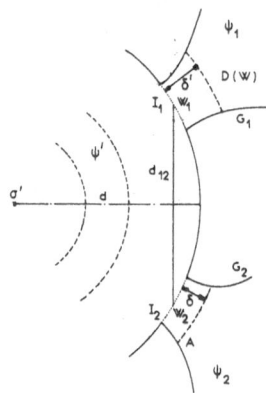

Fig. 1.

Preparations (*Figure* 1). A non-monochromatic and low intensity intermittent source σ emits microsystems S. At a distance d from σ is placed a spherical screen S, of radius d, centred on σ. Two circular windows W_1 and W_2 are cut out of S. The distance d_{12} which separates the centres of W_1, W_2 can be chosen arbitrarily big by increasing d. At the right of S the windows W_1, W_2 are continued by widening walls playing the role of guides G_1, G_2 (Figure 1). In these conditions each individual system S emitted by σ is described by quantum mechanics, at the left of S, by a spherical wave packet ψ', the front of which reaches at some given moment t, *simultaneously*, both windows W_1, W_2. From that moment on, at the right of S quantum mechanics describes the considered *one* system S by the superposition $\psi = (I/\sqrt{2})\psi_1 + (I/\sqrt{2})\psi_2$ of the two packets ψ_1 and ψ_2 transmitted respectively by the two windows W_1 and W_2. Because of the guides G_1, G_2 the supports I_1, I_2 of ψ_1, ψ_2, are finite, disjoint, and separated by the arbitrary distance d_{12}. Thus at the right of S one has prepared a superposition state of the type utilized in the proofs of Theorem 2.

The state ψ_1 can be prepared out of $\psi = (I/\sqrt{2})\psi_1 + (I/\sqrt{2})\psi_2$, by introducing an absorbing wall inside the guide G_2, at some distance S at the right of the surface (virtual) of the window W_2 (Figure 1). The state ψ_2 can be prepared similarly.

First stage of experiment: verification of the quantum mechanical predictions for w, ψ_1, ψ_2 and ψ. The distribution (and mean value) of w is measured separately in ψ (W_1, W_2 both open), ψ_1 (W_2 constantly shut) and ψ_2 (W_1 constantly shut). The results are compared in order to see whether the quantum mechanical *non*-additivity of the w-spectrum in ψ, with respect to the w-spectra in ψ_1 and ψ_2, is true or not. The problem, in this stage, is to define the theoretical conditions of observability of the sum of the 'interference terms' from the right side of (22) (interference in the w-*space*, even though in the physical space ψ_1, ψ_2 have disjoint supports), and to define a procedure which insures an w-resolution permitting the registration of the w-interference distribution, if it really exists. If in such appropriate conditions the predicted interference w-distribution is not registered, the quantum mechanical prediction from the right side of (22) is not *true* so that the non-locality Theorem 2 is not true either, so that a further stage of locality investigations is irrelevant. If on the contrary the w-interference term is observed, the following stage is relevant:

Second stage: locality investigation. For each system S emitted by σ the superposition state ψ is first prepared (W_1, W_2 both open). Then the preparation of ψ_1 out of ψ is started at a moment t_i by help of an absorbing shutter A dropped inside the guide G_2, at a distance ∂ from the surface of the window W_2. An w-measuring device $D(w)$ is placed inside the guide G_1 at a distance $\partial' = \partial + \varepsilon$ from the surface of the window W_1, ε being very small but sufficient for ensuring that when the front of the wave of the system reaches the level $\partial + \varepsilon$ the preparation of ψ_1 has already been accomplished (the term $(I/\sqrt{2})\psi_2$ of ψ has been suppressed) so that it is the wave-packet ψ_1 which reaches the w-measuring device $D(w)$. The condition

$$(21)' \qquad \frac{\varepsilon}{V_{\psi_1}} + \varDelta t < \frac{d_{12}}{C}$$

is required, where V_{ψ_1} is the group velocity for ψ_1 (depending on the mean energy chosen for the systems emitted by σ), $\varDelta t$ and c being defined by (21). The device $D(w)$ and the absorbing shutter A are each time set in action simultaneously and $D(w)$ is each time disconnected after a time inferior to d_{12}/C. If by repetition of this procedure the recorded w-distribution is identical with that found for ψ_1 alone in the first stage of experiment (i.e. if the w-interference term, supposed to have been previously found for ψ, is suppressed by the action of the shutter A) then it has to be concluded that either non-local effects have gone from I_1 to I_2, or the object named one microsystem S somehow is not confined on the support of the quantum mechanical state vector associated with its state. The problem to be solved for this stage is to realize the condition (21)' while furthermore ensuring, as in the first stage, conditions of observability of w-interference fringes (in the w-space).

Any observable w for which (20) is fulfilled can be envisaged, spin-components included. Upon a more detailed analysis the spin-component along the direction *perpendicular* on d_{12} might appear to be the most convenient choice. For the moment, however, we reserve our opinion concerning both the choice of w and that of the measurement procedure. If these choices raise questions and seem queer, this is a reflection from the queerness of the quantum mechanical theory of measurement. We believe that this queerness should not be allowed to act as an obstacle to any attempt of *verifying* the concepts and predictions of the orthodox theory of measurement.

4. REMARKS ON THE SUPERPOSITION STATES

Throughout the preceding study, the superposition states have played an essential role which draws attention on them.

There exists a tendency for confounding the superposition states with the mathematical decompositions permitted by the expansion postulate. This tendency has its source in the fact that the quantum mechanical formalism prescribes the same algorism for the calculation of predictions concerning a superposition state or concerning a mathematical expansion. However, quantum mechanics does distinguish – by their definitions – the superposition states from the mathematical decompositions. When this distinction is explicitly taken into account and then confronted with the identity of the algorisms prescribed for calculating the predictions, reasons appear for doubting the truth of certain predictions concerning superposition states. Indeed:

A quantum mechanical state vector ψ is defined at any time by the specification of boundary conditions B which determine an 'initial' form $\psi(q, t_0)$, and of an evolution operator H which determines the transform of $\psi(q, t_0)$ by the passage of time. We shall then write symbolically $\psi = \psi(B, H)$. The physical realization of both B and H is necessary for the physical realization of $\psi(B, H)$.

Let us now adopt the Schrödinger representation:

In a superposition state $\psi = a\psi_1 + b\psi_2$, the boundary conditions are different for ψ, ψ_1, ψ_2, while H can be the same, or not. To take an example, we suppose that H is the same. Then we write $\psi(B_1 + B_2, H) = a\psi_1(B_1, H) + b\psi_2(B_2, H)$, with a, b complex *constants* and ψ, ψ_1, ψ_2 having a time evolution corresponding to H. When ψ_1, ψ_2 'interfere' in the physical space or in some other w-space, this interference concerns two different states both *realized* simultaneously.

Consider now a mathematical decomposition of a state ψ, according to the eigenstates ϕ_{Q_i} of a dynamical quantity Q, $\psi = \Sigma_i c_i^\psi \phi_{Qi}$, such as is permitted by the expansion postulate. Boundary conditions B and a hamiltonian H are realized *only* for ψ. The ϕ_{Q_i} are constant vectors and the c_i^ψ are complex numbers *depending on B and H* via the definition

$$c_i^\psi(t_0) = \int \psi(t_0)\phi_{Q_i}\, dq, \qquad c_i^\psi(t) = \int \psi(t)\phi_{Q_i}\, dq.$$

Then we have to write $\psi(B, H) = \Sigma_i c_i^\psi(B, H)\phi_{Q_i}$. Here only ψ is physically realized, while on the right side of the equality $\psi(B, H)$ is *represented in terms of the standard state vectors ϕ_{Q_i} conceived, but not realized physically.*

When the probability law for some quantity $Q' \neq Q$ is calculated by the *same* algorism

$$\left| \int \psi \phi_{Q'_j}^* \, dq \right|^2 = \left| \int (a\psi_1 + b\psi_2) \phi_{Q'_j}^* \, dq \right|^2$$

or

$$\left| \int \psi \phi_{Q'_j}^* \, dq \right|^2 = \left| \int (\Sigma_i c_i^\psi \phi_{Q_i}) \phi_{Q'_j}^* \, dq \right|^2 ,$$

applied indistinctly to the superposition $\psi(B_1 + B_2, H) = a\psi_1(B_1, H) + b\psi_2(B_2, H)$ or to the expansion $\psi(B, H) = \Sigma_i c_i^\psi(B, H)\phi_{Q_i}$, this might involve erroneous identifications of statistics of real interactions between physically realized states, with mathematical interferences of standard states, conceived but not physically realized. Therefore we envisage that *the superposition principle might introduce certain false predictions.*

5. CONCLUSION

We have shown that Wigner's proof does not invalidate the concept of a joint probability of the position and the momentum variables, but raises instead a locality problem inside the one-system formalism of quantum mechanics.

In a critical research it might be illuminating to examine in detail a counter-example to a general assertion, instead of using it merely as a sufficient basis for the global rejection of this assertion. In constructive attempts the aim is the perception of some maximally unifying essence, and the choice of the maximal generality in the formulations ensures indeed a progression towards this aim. But in a critical attempt, on the contrary, the progress often lies in the identification of some particular circumstance of which a previous constructive effort has remained unaware, and which has therefore been erroneously forced into a conceptual structure imperfectly fitted for it, where its presence introduces distortions. Thus, in the present case, the connections established between the one-system locality problem and the particular type of state vectors for which this problem arises, suggest that the quantum theory might have erroneously integrated the description of momentum-dependent distributions, for these particular states. While the study of the position distribution for states $\psi = a\psi_1 + b\psi_2$ with $I_1 \cap I_2 \neq \emptyset$ has contributed to lead towards quantum mechanics, the study of the momentum- or spin-dependent distributions for the states $\psi = a\psi_1 + b\psi_2 -$ with $I_1 \cap I_2 = \emptyset$ or $I_1 \cap I_2 \neq \emptyset -$ might contribute to lead beyond the bounds of quantum mechanics.

ACKNOWLEDGEMENTS

I am profoundly indebted to Professor E. Wigner for having accepted to discuss a primitive version of the first part of this work. The theorem of non-locality, in its present form, has been suggested by Professor A. Shimony, with whom I had very stimulating exchanges of ideas. I am grateful to Dr J. S. Bell for important remarks on the possibility of an experimental study for the momentum. The ideas exposed here have much profited from numerous discussions with Dr D. Evrard and Dr F. Thieffine, to whom I am also indebted for having read the manuscript. Finally, I am indebted to Dr. G. Lochak for having read the manuscript in order to confront it with his own ideas on locality [17].

Laboratoire de Mécanique Quantique,
Faculté des Sciences de Reims,
Reims, France

NOTES

[1] The de Broglie–Bohm functional of this type is unnecessarily a priori restricted and thereby it introduces distortions[11].

[2] In a preceding work[11] it has been shown that the very *weak* hypothesis according to which the object denominated *one* microsystem cannot progressively extend over a spatial domain indefinitely increasing, suffices to entail a position distribution which in certain states is not rigorously identical with the quantum mechanical one. But the rigorous truth of the quantum mechanical prediction in this case has never been verified.

[3] Discussions on this subject with Dr D. Evrard have strongly contributed to the formation of this attitude.

[4] It seems interesting to compare our considerations on pp. 122–123, with the very pertinent analyses of Belinfante [16] on the hidden variables attempts made up to now.

BIBLIOGRAPHY

[1] Wigner, in *Essays in Honor of Alfred Lande*, The MIT Press, 1971.
[2] Bell, *Physics* 1 (1964), 195.

[3] Feynman, *Quantum Mechanics and Path Integral*, McGraw Hill, 1965.
[4] Evrard, Ph.D. Thesis, Université de Reims (France), May 1977.
[5] Park and Margenau, *Int. J. Theor. Phys.* 1 (1968), 211.
[6] von Neumann, *Mathematical Foundations of Quantum Mechanics*, Princeton University Press, 1955.
[7] Kochen and Specker, *J. Math. Mech.* 17 (1967), 59.
[8] Moyal, *Proc. Camb. Phil. Soc.* 45 (1949), 124.
[9] Bass, *C.R. Acad. Sci.* 221 (1945), 46.
[10] Cohen, *J. Math. Phys.* 7 (1966), 781.
[11] Mugur-Schächter, Evrard, and Thieffine, *Phys. Rev. D* 6 (1972), 3397.
[12] Mugur-Schächter, to be published.
[13] Mugur-Schächter, *C.R. Acad. Sci.* 266 (1968), 1053.
[14] Mugur-Schächter, *Annales de la Fondation Louis de Broglie* 1 (1976), 94.
[15] Mugur-Schächter, *Epistemological Letters*, Assoc. F. Gonseth, 1976.
[16] Belinfante, *A Survey of Hidden Variables Theories*, Pergamon Press, 1973.
[17] Lochak, *Quantum Mechanics, A Half Century Later*, this book, p. 245.

B. D'ESPAGNAT*

ON PROPOSITIONS AND PHYSICAL SYSTEMS[†]

ABSTRACT. It is pointed out that some already known inequalities (Bell's inequalities) and some new ones presented here can be used to test experimentally the validity of a general conception of the foundations of microphysics. This conception mainly consists in considering sets of propositions (having the structure of lattices but possibly of non-Boolean ones) and in assuming that when a proposition is true on a system S this constitutes an intrinsic property of S. It is shown that the results of some recent experiments corroborating the quantum mechanical predictions can be used to invalidate directly the general conception just described. This is done without reference to the general principles of quantum mechanics. More generally, our derivation does not depend for its validity on assuming the truth of any particular physical theory abstracted by induction from experimental knowledge.

1. INTRODUCTION

It is a truism that the advent of the modern physical theories – relativity, quantum mechanics, quantum electrodynamics, S matrix theory and so on – has induced us to abandon many familiar intuitive concepts. When we are asked why, our standard – and quite appropriate – answer is that one may well be skeptical about the possibility and usefulness of building up some alternative theoretical framework that (i) would incorporate and use these old concepts and (ii) would be as successful as each of our present-day theories in all their respective domains.

On the other hand, a motivated skepticism is far from being equivalent to a disproof. All the successful theories mentioned above are built upon elaborate sets of axioms that are justified only a posteriori, i.e. by the agreement between some of their consequences and observed facts (and by the absence of discrepancies). But it should be remembered that the appearance of two or more theories using very different basic concepts and yet accounting equally well for a given set of experimental data is *not* quite a rare event in physics. Hence the mere *existence* of the successful theories referred to does not establish that such and such a concept (or general view, or

*Postal address: Université de Paris Sud, Bât. 211, 91405 Orsay, France.

J. Leite Lopes and M. Paty (eds.), Quantum Mechanics, a Half Century Later, 147–169. All Rights Reserved
Copyright © 1975 by the Physical Review journal

the like) which they reject is indeed to be discarded once and for all, as being definitely inadequate. For that reason it is quite often asserted that in such a domain we cannot make any absolute statement. Quite frequently, it is even stated as an *obvious* truth that the judgements we can form on these matters are all dependent not only on the facts *but also on the general axioms of the existing theories.*

Still, if not for our *practice* of physical research, at least for our *understanding* of the whole subject we would like to know for certain as many items as we can concerning the adequacy or inadequacy of given concepts or general ideas. In particular we would be satisfied if we could, about some given concept or idea, establish, not only that it is *useless* at present (i.e. within the framework of the present-day theories) but that it is *false* in that it leads unavoidably to a contradiction with the data.

For that purpose, we stress again that a mere reference to the existing theories is not enough. How then should we proceed? Obviously by trying, as much as possible, to shortcircuit these theories. By trying to compare directly – or as directly as we can – the concept or idea with the experimental facts.

Now if our purpose is really to study only *one* concept or *one* particular idea – in isolation so to speak – then the above program is probably overambitious and cannot in fact be fulfilled. But at least it can be applied, as we show below, to a given *set* of concepts and ideas (assumptions). The result of course is weaker, since when we have shown that this set of concepts and assumptions directly contradicts known facts we can only conclude that one or more of these concepts or assumptions must be rejected, without being able to specify which one. Still, if this set contains only notions and ideas that are *all* deeply ingrained in our minds even this weaker result is interesting.

In this paper a set of concepts and assumptions is introduced (Section 2) that is already considered by most experts as not being compatible with quantum mechanics, at least in its most commonly accepted interpretations. The question we are interested in is: are we *sure* that this set is absolutely unacceptable, i.e. that it will remain so in *any* future theory or interpretation thereof? It is shown that we can answer that question positively provided only that we accept the validity of a few recent experimental results.

While the proof of the above statement is the main purpose of this paper, a subsidiary one is to supplement Bell's inequalities [1, 2, 3] with some new ones, and also to show that, when applied to special correlation effects, all these inequalities hold within a theoretical framework that is considerably larger than the one of the hidden variable theories. In particular a by-product of our method for deriving such inequalities is to make it clear that they have an already known large domain of application, which covers indeed the whole field of what can be called the 'classical' probabilities (by this expression we mean the set of all the theories in which elementary events (i) exist and (ii) are all such that several propositions can be formulated that are each necessarily true or false when applied to them). But it should be stressed that the main purpose of this article is definitely not to establish new inequalities nor indeed to put forward any new physical result. Rather it is to study a new *problem*, which consists in ascertaining directly whether or not a given set of general ideas and concepts is compatible with known facts, independently of *any* formalism.

Some of the views presented here were already put forward – in a provisional form – by the author at the 1972 Trieste Conference on the Physicist's Conception of Nature [4]. They are reformulated here since they fit naturally with the context.

2. CONCEPTS AND ASSUMPTIONS

The set of concepts and assumptions that we want to falsify directly – without reference to quantum mechanics – is the following one.

As regards the *concepts* we merely assume that we can use the words *system, isolated system* and *proposition* in the usual way. A silver atom is a system of a given type. A voltmeter, an electron are systems of other types. Propositions are defined operationally. We define a proposition **a** pertaining to a type T of systems S by specifying the instruments of measurement corresponding to it. We also define the orthogonal complement **a'** of **a** by specifying that it corresponds to the same instrumental device as **a** and that its measurement is said to give the value *yes* whenever that of **a** gives *no*, and conversely.[1]

These *concepts* cannot be completely separated from the *assump-*

tions that follow (to some extent, the distinction between assumptions and concepts is artificial in this context).

One of the ideas concerning physical systems that is most deeply ingrained in our general conceptions about Nature is that in some cases at least, some propositions are true about these systems: and that when it is the case it is so even if nobody is actually going to try to become conscious of the fact. Let us formulate precisely this idea in the following way.

ASSUMPTION 1. It is meaningful to associate to any proposition **a** defined on a type T of systems a family $F(\mathbf{a})$ of systems S of the type T, $F(\mathbf{a})$ being defined by the two following conditions: (i) the systems S that belong to $F(\mathbf{a})$ are those and only those that are such that if **a** were measured on S by any method the result *yes* would necessarily be obtained and (ii) the fact that a given S belongs to $F(\mathbf{a})$ is an intrinsic property of S (i.e. it does not depend on whether or not S will interact with some instrument devised so as to measure **a**).

Remark 1. Assumption 1 apparently conflicts with some at least of the conventional interpretations of quantum mechanics. In particular, it seems difficult to reconcile it with some of the views of the Copenhagen school concerning the role of the instruments and the inseparable wholeness they are supposed to constitute with the object. On the other hand this particular aspect of the Copenhagen interpretation has always remained somewhat controversial, even in the opinion of some physicists who consider themselves as being substantially in agreement with the conception of that school. Indeed, some of the latter physicists seem to have hoped to be able to restore the validity of our Assumption 1 by going to a non-Boolean logic[5], or to a non-Boolean calculus of proposition[6]. One of the points we expect to make in this paper is that such hopes cannot be maintained; and that this is true quite independently of any theory.

Remark 2. The possibility that systems of type T should exist that belong neither to $F(\mathbf{a})$ nor to $F(\mathbf{a}')$ is clearly not excluded by Assumption 1; nor is even the possibility that some systems should belong to no family of that sort at all. In particular we do not assume that if a is not true it is false. Indeed we do not even define a meaning for the latter epithet applied to a proposition bearing on a system.

Remark 3. No determinism – neither manifest nor hidden – is postulated.

DEFINITION 1.   Iff S belongs to $F(\mathbf{a})$, $\mathbf{a}$ is said to be *true* on S.

DEFINITION 2.   Let a system S be isolated between times t_a and t_b. $\mathbf{a}$ is said to be *persistent* on S between t_a and t_b iff the condition that $\mathbf{a}$ is true at time t_1 entails that it is true at time t_2, for any t_1 and t_2 satisfying

$$t_a < t_1 < t_2 < t_b.$$

ASSUMPTION 2.   Let $t_a < t_1 < t_2 < t_b$, let S be isolated between t_a and t_b and let $\mathbf{a}$ be persistent between t_a and t_b. Then if $\mathbf{a}$ is true at t_2 it is also true at t_1.

Remark 4.   Assumption 2 is again one of those that seem to be incompatible with at least some interpretations of quantum mechanics, although this, again is controversial. But anyhow it is an assumption that seems quite natural in view of Definition 2 and of our general opinion that in such matters some kind of time reversal principle should hold.

ASSUMPTION 3.   The fact that a proposition $\mathbf{a}$ is true on a system S at a time t cannot be changed by modifying the experimental devices with which S (or any other system) will interact at times posterior to t.

ASSUMPTION 4.   If $\mathbf{a}$ is true on S, then it is also true on any system $S + S'$ of which S is a part. Conversely if $\mathbf{a}$ is a proposition defined on systems of the type of S, if it bears on S and if it is true on $S + S'$, it is true on S.

3. CONSEQUENCES

Let us consider the experiment discussed by Bohm[7], Bell[1] and others[8]. A spin zero particle decays at time t_a into two particles U and V of equal spin S by means of a spin-conserving interaction. Let $\{\mathbf{e}_i\}$ be unit vectors defining directions in space. Let v_i be the proposition '$S^{(V)}(\mathbf{e}_i) = m$' and let u_i be the proposition '$S^{(U)}(\mathbf{e}_i) = -m$', where $S^{(W)}(\mathbf{e}_i)$ is the projection along $\mathbf{e}_i$ of the spin of particle W ($W = U$ or V). Propositions u_i and v_i can be defined by means of suitably oriented Stern-Gerlach devices. It is then apparent that v_i' is the proposition '$S^{(V)}(\mathbf{e}_i) \neq m$' and similarly for u_i'. On the other hand, if we did measure u_i and v_i, in any order we would always get either two answers *yes* or two answers *no*. This can be considered as a definition of the statement that the composite system $U + V$ has total spin zero (all measurements are assumed here to be 'ideal') and we can consider it as an experimental fact that systems $U + V$ prepared as stated above *do* have spin zero. Combined with Assumption 1, the fact that upon measurement of u_i and v_i we would certainly get either two *yes* or two *noes* implies that if the composite system $U + V$ belongs to $F(u_i)$ it also belongs to $F(v_i)$ and conversely.

Let us consider the case in which – at a time $t_2 > t_a$ – u_i is measured on U by means of some instrument A. Let us assume first that the result *yes* is obtained. Then, for the reason already mentioned it can be stated with certainty that a measurement of v_i on the corresponding system V would also give the result yes. According to Assumption 1, V therefore belongs – after time t_2 – to family $F(v_i)$. Since v_i is a persistent proposition on V from $t = t_a$ to $t = \infty$, Assumption 2 has then the consequence that that particular V belongs to $F(v_i)$ also at any time t_1 satisfying $t_a < t_1 < t_2$. Assumption 4 then shows that also the composite system $U + V$ of which the considered V is a part, belongs to $F(v_i)$. Because of the strict spin correlation established at time t_a it thus also belongs to $F(u_i)$.

Let us now assume that the result of the measurement made on U at time t_2 is *no*. Exactly the same argumentation then leads us unavoidably to the conclusion that in that case the composite $U + V$ system belongs at time t_1 to families $F(u_i')$ and $F(v_i')$.

Instead of considering one composite system $U + V$ only let us now consider N such systems, all identically prepared and all of them subjected to a measurement of u_i at t_2. For each of them the result of that measurement is necessarily *yes* or *no* so that each of these systems necessarily falls into one of the cases considered above. The previous argument therefore shows that at time t_1, under the conditions of the experiment and if Assumptions 1, 2 and 4 are correct, the composite systems $U + V$ all belong *either* to $F(u_i)$ and $F(v_i)$ *or* to $F(u_i')$ and $F(v_i')$. If now we take also Assumption 3 into account we must conclude that this situation *would also hold* if the measurement hitherto assumed to be made on U at time t_2 were *not* made at all, or were replaced by some other one. But then the same argumentation can be repeated over again with reference to a new pair u_j, v_j of propositions. Hence the conclusion is that in the special case of the decay considered here we have to deal with a situation in which it so happens that any composite $U + V$ system:

    (i) must belong either to $F(v_i)$ or to $F(v_i')$,

    (ii) must belong also to $F(u_i)$ in the first case and to $F(u_i')$ in the second one and

    (iii) belongs as a matter of fact to an infinity of such families at the same time since e_i can be chosen in an infinity of ways. We may question these conclusions but the point is that we may not do this without giving up one or several of Assumptions 1, 2, 3 and 4.

Remark 1.   This argumentation closely parallels the one developed by Einstein, Podolsky and Rosen[9] in order to show that quantum mechanics is incomplete. But it is used here with somewhat different assumptions and for a different purpose, since our objective is *not* to test any assumption (e.g. completeness) concerning the axioms of quantum mechanics. As a consequence – in contradistinction with what was the case as regards the article quoted above – the results obtained in this section do not yet constitute a difficulty as regards the assumptions we want to test since no contradiction exists between them and the experimental facts that are used here as reference. In particular, they are fully compatible with the experimental facts usually described under the headings 'the spin components along different directions are not simultaneously measurable'. Admittedly, the results in question imply for instance that *if* u_i were measured at t_1 on some system U the answer yes *would* be obtained and that *if* u_j were measured *instead* on the same U the answer yes *would* also be obtained. But it asserts nothing about any *actual* sequence of such measurements (concerning which the problem of the perturbation created by the first instrument would have to be taken into account) and, what is even more significant, it does not give us any operational means for effectively sorting out from the statistical ensemble a system U possessing these features. Indeed, under these circumstances it would even seem at first sight that the special character endowed to the considered composite systems $U + V$ by our assumptions has no observable implication whatsoever. If this conclusion were correct it would reinforce the view that sets of assumptions of this sort are 'legitimate but metaphysical'. But as we show below, a complete elucidation of the bearing of the Bell type inequalities must lead us – on the contrary – to give up this view since such inequalities (i) *can* be falsified and (ii) *are* consequences of the results derived directly in the present section from the considered set of assumptions.

Remark 2.   Some formulations (see e.g. [4]) introduce the notion of *atomic propositions.* When a is atomic then, if x is a proposition

$$\emptyset \subseteq x \subseteq a \Rightarrow x = \emptyset \qquad \text{or} \quad x = a,$$

it might seem that the results of this section preclude the possibilities of u_i or v_i being atomic on U and V respectively since the assertion $x = 'u_i$ and u_j' (which was shown to hold on some U's) entails u_i while being different from $\emptyset$. But the conclusion does not follow since – as pointed out in the foregoing remark – assertion x is not operational and therefore is not a proposition.

On the other hand this makes clear a point that could be important for the development of the theories gathered under the names of 'quantum logic' or 'quantum calculus of propositions'. This point is that any such theory that implicitly or not makes use of our set of assumptions implicitly contains 'built in' significant assertions – such as x above – that are different from propositions.

4. INEQUALITIES

The semi-positive definite character of the probabilities (that they cannot be negative) has many consequences – some of which have perhaps not yet been completely exploited in particular in conjunction with strict correlation phenomena. Here we derive Bell's inequalities and some generalizations thereof as simple, nay almost trivial, consequences of that semi-positive definiteness (these inequalities consequently apply for a wide range of physical theories and phenomena, including macroscopic, classical ones).

Through the use of the concepts of measure, conditional probabilities and so on (and of the corresponding short-hand notations) the following derivations could easily be formulated in concise, abstract terms. However this would conceal, rather than reveal, their intrinsic simplicity and (what is more important) their corresponding generality. Let us instead use the very simple notion of number of systems in an ensemble. The number of elements in a statistical ensemble is an inherently non-negative quantity; and the number of elements of the union of two disjoint ensembles is the sum of the numbers of elements of the two constituents. These two trivial but indisputable statements are essentially all we need and by formulating them in such a concrete manner we hope to show in a convincing way that the basis of the following deduction is extremely difficult to reject.

Let us then consider an ensemble E of system V of a given type T. Let $\{v_1 \ldots v_i \ldots v_n\}$ ($n \geq 3$) be a set of propositions – $\{v_1' \ldots v_i' \ldots v_n'\}$ being the set of their orthogonal complements – defined on systems of type T and such that every element of E belongs for any value of the index i, either to $F(v_i)$ or to $F(v_i')$, $F(v_i)$ and $F(v_i')$ being the families of systems defined in Assumption 1. In classical physics, ensembles E satisfying such conditions can be constructed in an extremely wide variety of cases (as already mentioned in the introduction). But even when propositions of a type more general than the classical ones are considered, it may happen (in particular cases) that such ensembles can be considered also. An example is provided by the ensemble $E = E_V$ of the systems V considered in the foregoing section. This, as shown in Section 3, is a consequence of the set of assumptions introduced in Section 2. Hence the following considerations apply also to E_V as soon as Assumptions 1 to 4 of Section 2 are made, which we assume to be the case.

Let us choose an approach originally used by Wigner[2] in order to deal with the hidden variables problem: to each element V of E let us associate a sequence $\sigma_1 \ldots \sigma_i \ldots \sigma_n$ of dichotomic quantities σ_i which have the values +1 (denoted +) if V belongs to $F(v_i)$ and −1 (denoted −) if V belongs to $F(v_i')$. Let us first consider three v_i only and let then

$$n(\sigma_1, \sigma_2, \sigma_3)$$

be the number of systems V in E that have the specified values of σ_1, σ_2, σ_3. Although we cannot know n it has a well defined value according to our assumptions (supplemented with the considerations of Section 3) in all the cases we consider. Moreover in all these cases

$$\sum_{\sigma_1, \sigma_2, \sigma_3} n(\sigma_1, \sigma_2, \sigma_3) = N,$$

where N is the total number of elements of E. ($N \to \infty$).

Let $M(i, j)$ $(i, j = 1, 2, 3)$ be the mean value on E of the product $\sigma_i \sigma_j$, so that, of course

$$-1 \leqslant M(i, j) \leqslant 1$$

and

(1) $$M(i, j) = N^{-1} \sum_{\sigma_1, \sigma_2, \sigma_3} \sigma_i \sigma_j n(\sigma_1, \sigma_2, \sigma_3).$$

PROPOSITION 1

(2) $$|M(i, j) - M(j, k)| \leqslant 1 - M(k, i), \quad \text{for} \quad i \neq j \neq k.$$

Proof. The quantity:

(3) $$M(i, j) - M(j, k) = N^{-1} \sum_{\sigma_1, \sigma_2, \sigma_3} \sigma_j (\sigma_i - \sigma_k) n(\sigma_1, \sigma_2, \sigma_3)$$

contains no term with $\sigma_i = \sigma_k$, hence only terms with $\sigma_k = -\sigma_i$, and can therefore be rewritten as

(4) $$M(i, j) - M(j, k) = 2N^{-1} \sum_{\sigma_1, \sigma_2, \sigma_3}{}' \sigma_j \sigma_i n(\sigma_1, \sigma_2, \sigma_3),$$

the symbol Σ' meaning that all the terms in which $\sigma_k = \sigma_i$ must be excluded from the summation and only these. Similarly

(5) $$1 - M(i, k) = N^{-1} \sum_{\sigma_1, \sigma_2, \sigma_3} (1 - \sigma_i \sigma_k) n(\sigma_1, \sigma_2, \sigma_3)$$

also contains no term with $\sigma_k = \sigma_i$ and can be rewritten – with the same convention – as

(6) $$1 - M(i, k) = 2N^{-1} \sum{}' n(\sigma_1, \sigma_2, \sigma_3).$$

In Equations (4) and (6) the summations bear on the same terms but in (6) all these terms are positive whereas in (4) some of them can be negative. Hence (2) follows, q.e.d.

Since any composite system $U + V$ that belongs to $F(v_i)$ also belongs to $F(u_i)$ as we have shown, $M(i, j)$ can be known experimentally. Indeed, in the $S = \frac{1}{2}$ case:

(7) $$M(i, j) = -P(i, j),$$

where $P(i, j)$ is the mean value of the (observable) product of $S_{e_i}^{(U)}$ and $S_{e_j}^{(V)}$. Equation (2) therefore gives rise to Bell's inequalities[1]:

(8) $\qquad |P(i, j) - P(j, k)| \leqslant 1 + P(k, i).$

PROPOSITION 2

(9) $\qquad M(12) + M(23) + M(31) \geqslant -1.$

Proof. The left hand side – which we designate by A – can be written as

(10) $\qquad A = N^{-1} \sum_{\sigma_1, \sigma_2, \sigma_3} (\sigma_1\sigma_2 + \sigma_2\sigma_3 + \sigma_3\sigma_1)n(\sigma_1, \sigma_2, \sigma_3),$

(11) $\qquad A = (2N)^{-1} \sum_{\sigma_1, \sigma_2, \sigma_3} [(\sigma_1 + \sigma_2 + \sigma_3)^2 - 3]n(\sigma_1, \sigma_2, \sigma_3).$

Since $\sigma_i = \pm 1$ the quantity inside square brackets can only take the values $+6$ (for $\sigma_1 = \sigma_2 = \sigma_3$) and -2 (otherwise). Hence

(12) $\qquad A \geqslant -N^{-1} \sum_{\sigma_1, \sigma_2, \sigma_3}'' n(\sigma_1, \sigma_2, \sigma_3),$

where Σ'' is a summation extended to all the terms for which not all three σ's are equal. Obviously $\Sigma'' \cdots \leqslant N$, and (9) follows. For the observable quantities $P(i, j)$, (9) gives the new inequality

(13) $\qquad P(12) + (P(23) + P(31) \leqslant 1.$

Remark 1. In the special but important case (used above as an example) in which U and V are equal spin particles in a state of zero total spin (and in the similar experiments using photons) it can be shown that (13) combined with (8) is equivalent to an inequality derived by Gutkowski and Masotto [10] and relating with one another not the P's but the corresponding numbers of systems (probabilities). On the other hand the G.M. inequality is based on the fact that the probabilities of the results $\sigma_i = \pm 1$ are equal. Experiments could probably be imagined in which such an equality would not hold, but for which (9) (or (13)) would still be valid.[2]

Remark 2. The effect of the strict spin correlation between U and V is two-fold. (i) Together with Assumptions 1 to 4 it integrates any ensemble of systems V to the class of those any element of which belongs either to $F(v_i)$ or to $F(v_i')$ and (ii) it has the effect that the quantities $M(i, j)$ become observable, by means of the $P(i, j)$.

PROPOSITION 3.   Let

$$(14)\quad\begin{aligned}Nx &= n(+, +, +) + n(-, -, -) \leqslant N,\\Ny &= n(+, +, -) + n(-, -, +) \leqslant N,\\Nz &= n(+, -, -) + n(-, +, +) \leqslant N,\\Nt &= n(+, -, +) + n(-, +, -) \leqslant N.\end{aligned}$$

Then

$$(15)\quad\begin{aligned}A &= \frac{1 - M(12)}{2} = z + t,\\B &= \frac{1 - M(13)}{2} = y + z,\\C &= \frac{1 - M(23)}{2} = y + t,\end{aligned}$$

or

$$y = (B + C - A)/2; \quad z = (A + B - C)/2;$$
$$t = (A + C - B)/2,$$

The only independent inequalities (or equalities) satisfied by x, y, z and t on these grounds are

(16) $x \geqslant 0, \quad y \geqslant 0, \quad z \geqslant 0, \quad t \geqslant 0,$

(17) $x + y + z + t = 1.$

Hence the only inequalities that the additive and the semi-positive definite nature of the entities 'numbers of systems' can generate for linear combinations of A, B, C are those derived from (16) and (17) by substitution. The three last inequalities (16) give inequalities (2) (Bell's inequalities). Equation (17) and the first inequality (16) give together inequality (9). The first inequality (16) gives no information. It follows that the inequalities (2) and (9) exhaust the list of the inequalities satisfied by linear combinations of the $M(i, j)$ as a consequence of additivity and semi-positive definiteness.

PROPOSITION 4.   Let us consider a fourth unit vector e_4 and the corresponding proposition v_4. Then

(18) $-2 \leqslant M(12) + M(13) + M(24) - M(34) \leqslant 2.$

Proof.   Let the symbols Σ' and Σ'' denote summations over the possible values of the σ's from which the terms having respectively $\sigma_2 = \sigma_3$ and $\sigma_2 = -\sigma_3$ are excluded. Let the middle term of (18) be denoted by B. With obvious notations, B can be written

$$B = N^{-1} \sum_{\sigma_1, \sigma_2, \sigma_3, \sigma_4} [\sigma_1(\sigma_2 + \sigma_3) + \sigma_4(\sigma_2 - \sigma_3)]n(\sigma_1, \sigma_2, \sigma_3, \sigma_4),$$

$$B = N^{-1}\left[2\sum{}'' \sigma_1\sigma_2 n(\sigma_1, \sigma_2, \sigma_3, \sigma_4) \right.$$
$$\left. + 2\sum{}' \sigma_4\sigma_2 n(\sigma_1, \sigma_2, \sigma_3, \sigma_4) \right].$$

Hence

$$-2N^{-1}\left(\sum{}'' n(\ldots) + \sum{}' n(\ldots) \right) \leqslant B$$

$$\leqslant 2N^{-1}\left(\sum{}'' n(\ldots) + \sum{}' n(\ldots) \right),$$

and therefore (since the ensembles Σ' and Σ'' are disjoint):

$$-2 \leqslant B \leqslant 2 \quad \text{q.e.d.}$$

Equation (18) and the inequalities derived by permuting the symbols give rise to the so-called 'generalized Bell's inequalities' between the $P(i, j)$. These inequalities were first derived within the hidden variables conception by Clauser, Holt, Horne and Shimony[3, 11]. Within that conception they hold true even if a strict correlation does not hold between U and V in the sense in which this concept is introduced in Section 3. On the contrary if we only assume the validity of the set of assumptions listed in Section 2 these inequalities are only valid in the cases in which strict correlations hold (i.e. in the case of a total spin zero in our example).

More generally, let us now consider the case in which m distinct propositions u_i are taken into account. Let us first consider the case in which m is odd. Then we have

PROPOSITION 5

(19) $\displaystyle\sum_{i<j} M(i,j) \geq \frac{1-m}{2}$, m odd,

and correspondingly

(20) $\displaystyle\sum_{i<j} P(i,j) \leq \frac{m-1}{2}$, m odd.

Proof.   The left hand side of (19) is

(21) $\displaystyle (2N)^{-1} \sum_{\sigma_1,\ldots,\sigma_m} [(\sigma_1 + \cdots + \sigma_m)^2 - m]n(\sigma,\ldots,\sigma_m)]$,

the smallest term among those written inside square brackets in (21) has value $1-m$. Equation (19) follows.

When the parity of m is not specified the inequality obtained by this method is less stringent. It is

(22) $\displaystyle \sum_{i<j} M(i,j) \geq -m/2$.

The corresponding inequality for the $P(i,j)$ is unlikely to be falsified by experiments made on the $U + V$ systems such as those considered in Section 3 since – contrary to (20) – it is always satisfied by spin $\frac{1}{2}$ systems obeying quantum mechanics. This follows from the fact that $P(i,j)$ can then be written

$$P_{\text{q.m.}}(i,j) = -(\mathbf{e}_i \cdot \mathbf{e}_j),$$

so that

(23) $\displaystyle \sum_{i<j} P_{\text{q.m.}}(i,j) = 2^{-1} \cdot [(\mathbf{e}_1 + \cdots + \mathbf{e}_m)^2 - m] \leq \frac{m}{2}$,

the equality being realized for $\mathbf{e}_1 + \cdots + \mathbf{e}_m = 0$.

On the other hand the derivation of inequality (22) can be applied to the more general case in which $M(i,j)$ is the mean value of the product $X_i X_j$ of two random variables. Denoting by σ_i the values taken by the X_i we have

$$\sum_{i<j} M(i,j) = N^{-1} \sum_{\sigma_1,\ldots,\sigma_m} \sum_{i<j} \sigma_i \sigma_j n(\sigma_1,\ldots,\sigma_m) = (2N)^{-1} \sum_{\sigma_1,\ldots,\sigma_m}$$
$$\times [(\sigma_1 + \cdots + \sigma_m)^2 - (\sigma_1^2 + \cdots + \sigma_m^2)]n(\sigma_1,\ldots,\sigma_m),$$

and hence

(24) $\sum_{i<j} M(i,j) \geq -\left(\sum_i M(i,i)\right)\Big/2; \quad i,j = 1,\ldots,m.$

In the case in which the X_i are centered and have equal root mean squares (24) reduces to the inequality

(25) $\sum_{i,j} r(i,j) \geq -m/2; \quad i,j = 1,\ldots,m,$

between the correlation coefficients $r_{ij} = M(i,j)(M(ii)\cdot M(jj))^{-\frac{1}{2}}$. Inequalities (24) and (25) belong essentially to ordinary probability theory and they should be used as such. Inequalities (2), (9), (18) and (19) can also be used within the same framework. When the X_i can be considered as constituting together a stationary stochastic function of the index i, inequality (25) reflects the well-known fact that any correlation function of such a stochastic function is positive-definite.

5. DISCUSSION

Let us carefully distinguish between on the one hand the verifiable predictions of quantum mechanics (which are unambiguous in every case) and on the other hand both its formalizations and its conceptual interpretations (which are varied and/or controversial). If we believe that the *verifiable predictions* are all correct, then the content of the present article forces upon us (with no commitment to a particular formalism, or to a particular interpretation) the conclusion that the set of the assumptions listed in Section 2 cannot be kept since such inequalities as (8), (13) and – more generally – (20), that follow from these assumptions, are violated in some cases by the said verifiable predictions. Such cases include those in which the systems U and V considered in Section 3 are spin $\frac{1}{2}$ particles and in which the unit vectors e_i are chosen in some special ways. This was shown by Bell[1] as regards inequalities (8). As regards inequalities (13) and (20) it follows for instance from the fact that the symmetrical configuration $\Sigma_i e_i = 0$ corresponds to the equality sign in relation (23) and hence to a violation of (20).

If we do *not* take it for granted that all the verifiable predictions from quantum mechanics are true then we must rely upon direct experiment. Fortunately, in connection with the hidden variables problem, experiments have been made[12, 13, 14, 15] that can serve to test inequalities (8) or their generalizations[3]. Others are in progress[16]. Unfortunately there seems to be – for the time being – some experimental discrepancies between the results. This is a supplementary motivation for varying the tests; and this can be achieved by using inequalities (20) also: for instance in their simplest version which is (13). But independently of that, it should be pointed out that the set of the experiments that are suitable for testing the hidden variables hypothesis does not coincide exactly with the set of the experiments that are suitable for testing the set of assumptions under discussion in the present article. For example, the hypothesis that hidden variables exist can be tested (by using the generalized Bell's inequalities already mentioned) even in the case in which the correlation between the spins of U and V is not strict in the sense in which this concept is used in Section 3 above. On the contrary, for testing the validity of the set of assumptions listed in Section 2 the strict correlation effect is essential. This can also be done with photons, as in Refs. [12, 13, 14 and 15] and if the results of Ref. [13] are taken at face value they certainly seem to contradict the consequences that we have derived from the said set.

It may be noted here that the condition that the spins of U and V should be $\frac{1}{2}$ or that U and V should be photons is by no means necessary for the validity of the considered tests.[3]

As a last remark bearing on experimentation let it be pointed out that when the considered strict correlation takes place between (pseudo) vectors – as in the example studied above – the experimental devices that serve in testing inequality (13) can be the same as those that are used for testing inequalities (8). This is a consequence of the fact that if the direction of one of the vectors e_i is inversed two of the $P(i, j)$ change sign and (13) becomes identical to one of the inequalities (8). It does *not* mean that (13) is equivalent to (8). For example, in the symmetrical configuration $e_1 + e_2 + e_3 = 0$ (13) is violated by the quantum-mechanical $P(i, j)$ while (8) is not. Nevertheless the fact just mentioned – together with inequality (13) proved above – has the straightforward consequence that only unoriented directions in space are important. Given three such directions, the question whether the system of inequalities (8) and (13) is violated or not by the data does not depend on the orientation chosen on any of these lines in order to label as 'yes' the response of the instrument oriented along this line.

When photons are used, then of course the inequalities (2), (9) and (19) – i.e. inequalities involving the M's – are to be used instead of those involving the P's; this being true if the conventional definition of the measured quantities, namely

$$(26) \qquad M(i, j) = N^{-1}(n(++) + n(--) - n(-+) + n(+-)),$$

is used, where + means 'pass the polarizer', – means 'fails to pass' and $n(\sigma_1, \sigma_2)$ are numbers of photon pairs. The transition from (9) to one of the Bell's inequalities (2) – as well as between the latter – then corresponds to exchanging 'pass' and 'fails to pass' for one of the two photons, that is to an invariance with respect to rotations of $\pi/2$ of the polarizers.

For the rest of the discussion let us assume as a working hypothesis that the experimental results have confirmed or will confirm the observable predictions of quantum mechanics, so that the set of assumptions of Section 2 is falsified. The questions are then (i) what does this imply as regards the existing approaches of quantum mechanics and (ii) more generally, by what other assumptions can we replace the set under discussion? On these two questions we formulate here but a few remarks.

(i) Question of the Approaches to Quantum Mechanics

As regards this point the main interest of the present analysis is probably that it discriminates between several interpretations of conventional quantum mechanics and that it questions some of them. In particular it discriminates between the Copenhagen formulations and at least some versions of what could be called for short the axiomatical-logical formulations. As mentioned in the introduction, the conclusions we have reached are in no disagreement whatsoever with the Copenhagen interpretation, simply because this interpretation does *not* postulate the entire set of assumptions that has been falsified above. Indeed that interpretation discards Assumption 1, since (as Bohr in particular has repeatedly stressed) according to it, microsystems do not have any properties *of their own* (which means: that would be independent of the experimental arrangement, *including* the apparatus with which these microsystems *will* be observed).

On the other hand, some of the axiomatical-logical formulations *do* involve assumptions that are strictly equivalent to our Assumption 1. For the sake of definiteness let us for example consider the approach of Jauch and Piron[6]. These authors first define 'yes-no experiments' and decide to 'say that the yes-no experiment α is *true* if a measurement of α will give the result yes with certainty'. Then they accept it as an empirical fact that certain pairs α, β of such experiments have the property

(27) α true $\Rightarrow \beta$ true,

which they write $\alpha < \beta$. They call 'equivalent' two yes-no experiments α_1, α_2 having the properties that

(28) $\alpha_1 < \alpha_2$ and $\alpha_2 < \alpha_1$.

Next, they denote by $\mathbf{a} = \{\alpha\}$ the class of all such yes-no experiments and call $\mathbf{a}$ a *proposition*. Finally the quoted authors observe that $\mathbf{a}$ is true iff any (and therefore all) of the $\alpha \in \mathbf{a}$ are true and decide that 'if the proposition $\mathbf{a}$ is true we shall call it a *property* of the system'.

   Now, the question is: when we introduce in this way the notion of propositions defined on systems can we avoid making Assumption 1 of Section 2? It seems that the answer is no. Since the fact that $\mathbf{a}$ is true is considered as a property of certain systems we may call $F(\mathbf{a})$ a family of such systems; and then the fact that a given system S belongs to $F(\mathbf{a})$ is (this is a tautology!) an intrinsic property of S, independent of whether S will be observed by such or such an instrument. The only conceivable doubt we might have would be in connection with the definition of truth. As we have just seen, the quoted authors use the future tense for defining the truth of a yes-no experiment α (they write '*will* give the result yes') whereas in our Assumption 1 the conditional (*would*) was used. If this use of the future were to imply that α is true only in the cases in which an instrument for measuring α is actually set up (and ready for the measurement) then we would again have to do with a theory in which, as in Bohr's conception, the microsystems have no properties independent from the complete experimental environment: we would thus avoid making Assumption 1. Unfortunately, if we understand the use of the future instead of the conditional in such a restrictive way (i) we come in conflict with the sentence quoted above that $\mathbf{a}$ (and therefore also α) is 'a property of the system' and (ii) we get into difficulties in giving a meaning to Relation (27): if the statement 'α true' has a meaning only when a complete experimental device is present, that is designed so as to measure α in the future, then how can this statement imply 'β true', an assertion which is related to some other, quite different experimental arrangement?

As a result of this discussion we are tempted to believe that the very method by means of which the quoted authors introduce the concept of a proposition makes it impossible for them to avoid making effectively Assumption 1. On the other hand we also believe that this same method is entirely in the spirit of the general axiomatical-logical approach to the foundations of quantum mechanics and that, far from being unduly specific it has – on the contrary – the great merit of making explicit what was implicit before it. In particular we believe that the said axiomatical-logical approach is inherently based on the idea that, somehow, even micro-systems *have* properties of their own, albeit these properties are described by propositions not obeying the usual Boolean logic. But if so, then it is this entire approach the validity of which is questioned by the present analysis (unless it could be shown that it does not postulate the validity of Assumption 2, 3 and 4 or unless the very existence of strict correlations is doubted, see e.g. Jauch in Ref. [8]. But none of these possibilities seems likely).

(ii) *Question of the Alternative Assumptions*

Quite independently of the whole controversy that is still going on on the foundations of quantum mechanics it follows from the present analysis and from the experimental results[12, 13] (if taken at their face value) that we must abandon one at least of the Assumptions 1 to 4 of Section 2.

If the usual notion of a system is kept, it seems rather artificial to give up Assumption 4 only. Analyses of the Einstein, Podolsky and Rosen (E.P.R.) problem that seem to proceed along this line are occasionally put forward but as a rule they implicitly deny some other assumptions of the set also.

As regards Assumption 3, there exist some subtle ways of violating the principle it refers to while keeping all the other ones and producing no *observable* effects of the future on the past: it is well known that deterministic theories can be found that violate none of the predictions of quantum mechanics; they are of the contextualistic[17] non-separable variety. But their non-separability (i) leads to no observable violation of the principle of finite velocity propagation of signals and (ii) can be accounted for as a consequence of a retroactive effect: in the phenomenon described in Section 3 this effect would consist in a retroactive influence of the measurement made in U at time t_2 on the parameters describing the state of the system $U + V$ at time t_a. Up to now however the scientific community seems to be reluctant as regards the idea that the future could act upon the past, even in a way that is not directly observable.

The idea that Assumption 2 should be violated is somewhat less unattractive. At any rate it does not sound completely unfamiliar to many of the theorists who have studied the foundations of quantum mechanics. But the main question is: should it be abandoned *alone*? If it is, then a strange kind of irreversibility is thereby introduced in the fundamental laws of physics. Hence it seems more natural to abandon also Assumption 1.

Finally there are two main possible substitutions for Assumption 1. One of them introduces the idea of a non-separability existing between the micro-system and the experimental arrangement (including the instruments with which the system will *later* interact). This seems to have been Bohr's view and the essence of his answer to the EPR criticism. Along with many satisfactory aspects such a view has the well known but nevertheless surprising feature of expressing the laws of the micro-world by using approximate classical concepts referring essentially to *our* experience of the macro-world. Moreover, it also violates the general principle lying behind Assumption 3. The other possible substitution to Assumption 1 offers a way of avoiding this: it introduces a non-separability between micro-systems that have once interacted. That type of non-separability is closely parallel to the non-separability of the quantum-mechanical wave function. It is the (sometimes implicit) common feature of two or three otherwise different descriptions: the one that introduces hidden variables in a deterministic[18, 19] but contextualistic and non-separable theory, the one that makes consciousness an active agent[20, 21, 22] and finally – if it can be proved that it does not reduce to one of the latter two – the description that makes objective the entire wave function of the Universe[23, 24].

6. CONCLUSION

For the sake of convenience let us call *principle of separability* the following principle, as formulated by Einstein. If S_1 and S_2 are two systems that have interacted in the past but are now arbitrarily distant 'the real, factual situation of system S_1 does not depend on what is done with system S_2 which is spatially separated from the former' [25]. The result of the foregoing analysis is that the set of assumptions listed in Section 2 has been falsified directly by preliminary experiments (although these require confirmation) and that the false assumption in the set is probably Assumption 1. This in turn implies that the principle of separability is false.

Superficially such a conclusion seems neither surprising nor new. After all the non-locality – in the general case – of the many particles wave functions is quite obvious. It finds its best illustration in the Pauli principle (which also questions the possibility of individualizing systems in the way separability would have it). Nay, even classical physics admits of correlations between spatially separated events and hence of sudden changes of the probability distributions, induced by distant measurements. On the other hand the very easiness with which we find these apparent counter-examples to the separability principle should make us doubtful about their real validity as such. Obviously, none of the facts we have just listed were unknown to Einstein! That this latter author could nevertheless give credence to the separability principle should therefore induce us to try and be as critical in the use of our conceptual frameworks as we are accustomed to be in the use of our mathematical formalism. Now as soon as we decide to make such an effort we discover that of course Einstein was quite right. In his times, separability could be questioned *but could not be disproved*. For example, the fact alluded to above that distant correlations can take place even between spatially separated events (when influenced by some common anterior one) has – in fact – nothing to do with the principle of separability, which refers, as just recalled, not to our knowledge but to 'the real factual situation'. More generally, the non-locality of the many particles wave function cannot be used straightaway as an argument against separability. For that purpose it must indeed be associated with an *interpretation* of that wave function; and this leads to quite a long chain of arguments, that hinges on the validity of the general axioms of quantum mechanics and the meaning we give to them; and that has led to long and subtle controversies.

Since our whole analysis is completely independent from quantum mechanics its conclusion against separability is free from such inconveniences. Its main defect is that the experimental results which should normally constitute its firm basis are quite recent and are somewhat controversial. We may however be confident that the experimental ambiguities will be resolved very soon.

ACKNOWLEDGEMENTS

It is a pleasure to thank Dr. A. Frenkel for important remarks and suggestions.

Laboratoire de Physique Théorique et
Particules Elémentaires,
Orsay[4]

NOTES

[†] This text has already been published in *Physical Review D* **11** (1975), 1424, under the title 'Use of Inequalities for the Experimental Test of a General Conception of the Foundation of Microphysics'. We thank the editors of this journal for kindly permitting its reproduction in this series.
[1] Details concerning the notion of propositions as used here can be found in textbooks, see e.g. J. M. Jauch in Ref. [4].
[2] Note that $P(i, j') = -P(i, j)$ if $e_{j'} = -e_j$. When such relations are taken into account all the inequalities (8) and (13) can be deduced from just one of them, e.g. (13). Similarly inequalities (16) are then not independent. See Section 5.
[3] The $P(ij)$ defined as correlations between U and V satisfying Equation (7) are then still observable but they are no more equal to the mean values of $S_{e_i}^{(U)} \cdot S_{e_j}^{(V)}$.
[4] Laboratoire associé au Centre National de la Recherche Scientifique.

BIBLIOGRAPHY

[1] Bell, J. S., *Physics* **1** (1964), 195.
[2] Wigner, E. P., *American Journ. Of Physics* **38** (1970), 1005.
[3] Clauser, J. F., Horne, M. A., Shimony, A., and Holt, R. A., *Phys. Rev. Letters* **23** (1969), 880.
[4] Jauch, J. M., *Foundations of Quantum Mechanics*, Addison Wesley, Reading

Mass, U.S.A.; Mehra, J. (ed.), *The Physicist's Conception of Nature*, D. Reidel, Dordrecht, Holland.
[5] von Weizsäcker, C. F., *Natürwiss.* **42** (1955), 521.
[6] Jauch, J. M. and Piron, C., *Helv. Phys. Acta* **42** (1969), 842.
[7] Bohm, D., *Quantum Mechanics*, Prentice Hall, Englewood Cliffs N.J.
[8] *Foundations of Quantum Mechanics*, proceedings of the IL Enrico Fermi International Summer School, Acad. Press, New York (esp. articles by J. M. Jauch, J. S. Bell, L. Kasday, A. Shimony, E. Wigner, B. de Witt); B. d'Espagnat, *Conception de la Physique Contemporaine*, Hermann, Paris.
[9] Einstein, A., Podolsky, B., and Rosen, N., *Phys. Rev.* **47** (1935), 777.
[10] Gutkowski, D. and Masotto, G., Catania preprint.
[11] Bell, J. S., in Ref. [8].
[12] Kasday, L., in Ref. [8].
[13] Freedman, S. J. and Clauser, J. F., *Phys. Rev. Letters* **28** (1972), 938.
[14] Horne, M. A., Harward report.
[15] Faraci, G., Notarrigo, S., Pennisi, A., Gutkowski, D., *Boll. SIF* **93** (1972), 39.
[16] Lahemi, M. and Mittig, M., *Phys. Rev.* **14D** (1976), 2543.
[17] Shimony, A., in Ref. [8].
[18] de Broglie, L., *Journ. de Phys.* **5** (1927), 225.
[19] Bohm, D., *Phys. Rev.* **85** (1952), 166, 180; Bell, J. S., On the Hypothesis that the Schrödinger Equation is Exact', Contrib. to Penn. State Univ. Conference (Sept. 1971) Ref. TH 1424 CERN.
[20] von Neumann, J., *Mathematical Foundations of Quantum Mechanics* (Engl. translation), Princeton 1955.
[21] Wigner, E. P., *Remarks on the Mind Body Question* in the Scientist Speculates, London; article in Ref. [8].
[22] London, F. and Bauer, E., *la théorie de l'observation en mécanique quantique*, Hermann, Paris.
Costa de Beauregard, O., Boston preprint.
[23] Everett, H., *Rev. Mod. Phys.* **29** (1957), 454. de Witt, B., in Ref. [8].
[24] Cooper, L. N. and van Vechten, D., *Amer. Journ. Phys.* **37** (1969), 1212.
[25] Einstein, A., in P. Schilpp (ed.), *Albert Einstein Philosopher Scientist*, Lib. of Living Philosophers Evanston Ill.

JEAN-MARC LÉVY-LEBLOND*

TOWARDS A PROPER QUANTUM THEORY

(*Hints for a Recasting*)

> Up to now, philosophers have only interpreted quan-
> tum theory. The point, however, is to transform it.
> (after the 'Theses on Feuerbach').

INTRODUCTION

Fifty years old, and not yet grown-up!

Despite the festive character of this collection, aimed at celebrating
half a century of quantum mechanics, let me take the risk of asking
a few indecorous questions. The chronological reference of this book,
to start with, might be worth some considering. Could not we imagine
that analogous volumes, with similar titles, were or will be conceived
in Berlin in 1950, Zürich in 1955, Manchester in 1963, Göttingen in
1975, Cambridge in 1975 too, Vienna in 1976, etc., celebrating various
possible birthdays of quantum mechanics[1], comparable in importance to the present one[2]? In fact, we know well that *none* of
these dates by itself could fully symbolize the breaking forth of a new
physics. It would not be sufficient either to list the succession of these
dates, would it be in detail, to account for this emergence. History
cannot be reduced to chronology. Far from being a sequential
enumeration of events, a cumulative description of linear processes, it
requires a retroactive analysis, a critical point of view. The history of
sciences, as any history, cannot but be written in the present tense. In
other terms, the history of science itself has a history, as may be
proven by the title of this book; celebrating the jubilee of the
foundation of a '*wave* mechanics', we take into account, rightly but
implicitly, the practice of these past fifty years, in modifying a limited
and inadequate terminology to replace it by a more generally valid
one, so that we now speak of '*quantum* mechanics'. This sensible

* Postal address: Laboratoire de Physique Théorique, Université Paris VII (Tour
33–43, 1er étage), 2, Place Jussieu, 75221 – Paris Cédex 05.

J. Leite Lopes and M. Paty (eds.), Quantum Mechanics, a Half Century Later, 171–206. *All Rights Reserved*
Copyright © 1977 *by D. Reidel Publishing Company, Dordrecht-Holland*

unfaithfulness to the very work which motivated this celebration, may be understood from a general point of view. Indeed, the history of a scientific field does not close with the end of its springing up period. Quantum mechanics was established during the first quarter of this century, through a scientific activity sometimes considered as 'revolutionary'; one should not conclude, despite some appearances, that the following half-century, leading us at the present day, only saw a 'normal' activity, consisting of merely applying a 'paradigm' set up by the great masters.[1] By the very fact that any new physical theory is born in a difficult breaking off with the preceding ones, it still bears their stamp: as ever, the new for a long time shows the mark of the old. Well after the emergence and development of a new theory, there remains various contradictions between, on the one hand, its intrinsic structure and conceptualization (such as they keep appearing with an increasing clarity) and, on the other hand, the temporary forms it could not but borrow. I will describe below several examples of this phenomenon for the case of quantum mechanics. It is the effect of the experimental and theoretical prac- tices within the new field to 'transform it into itself', by progressive elimination of its archaic and irrelevant notions and formulations. This *recasting* process by no means is less important, historically speaking, than the more spectacular breaking off which precedes and allows it[4].

The importance here of these general considerations comes from the rather paradoxical situation of quantum theory in that respect. The most recent of the great theoretical syntheses of physics, this last-born child is a backward one. It looks as if the recasting process, as described above, had not really taken place for quantum physics, or, at least, had remained in a mostly implicit stage. That our Colloquium will spend much of its time debating some of the same basic epistemological problems that were already discussed fifty years ago, may be taken as evidence that very little recasting indeed has been achieved. A detailed study of most textbooks in quantum physics could yield another proof; the deeply repetitive character of these books with respect to one another, the absence of any moder- nizing in the terminology as well as in the description of the theoreti- cal structure or in the discussion of the fundamental concepts, ex- press, it seems to me, a state of sclerosis without precedent in the history of physics.

For I doubt that, if a Colloquium was held in 1915 to celebrate "half a century of electromagnetism"[2][5], it included discussions about the properties of the ether, the physical reality of the 'displacement current' (even though this execrable terminology has been maintained), or the interpretation of Hertz's experiments. Through these fifty years, a thorough recasting of electromagnetism had been achieved; the field concept had emerged, the spatio-temporal framework of the theory had been brought to light (if I dare say so) by Einsteinian relativity, the formulation of Maxwell's equations had been deepened and tightened. A comparison between Maxwell's first papers and textbooks of the twenties bears a clear testimony in that respect. Similar statements could be made for the good old 'classical' mechanics, or thermodynamics, etc. Of course, it must be emphasized that none of these recasting processes yet should be considered as closed; even though the domains of validity of such ancient physical theories may be well defined by now, their internal structure keeps modifying under the influence of the new theoretical syntheses which overtake and extend them. As an example, the role of symmetry principles and invariance considerations, come to the foreground of quantum theory (see below), has also taken a great importance in the reformulation of more 'classical' theories. At least, it can be said that, up to a recent past, these recasting processes in physics had not met with major obstacles.

What, then, is the nature of these obstacles which have maintained the recasting of quantum theory to a late, superficial and insufficient development? They derive, I believe, from the particular historical situation in which quantum physics was born. Two related features are of importance here: (1) the upsetting of scientific *practice* as a social activity, (2) the change in the philosophical (not to say ideological) conditions for the elaboration of *theory*. As to the first of these two topics, it is to be emphasized that the end of the first quarter of this century marks precisely enough the boundary between two modes of production of scientific knowledge. The ancient mode was essentially one of craftsmanship. It was based on the individual skill of scholars, working either in isolation or surrounded with a few pupils and students; hierarchical relationships were of the patronizing type, the values were devotion to progress, scientific integrity, humanitarian ascesis and ethics of knowledge (I am dealing here with 'values', that is with the latent or patent ideology of the scientific milieu, not with its real functioning, sometimes mean enough and in any case rather trite). After the First World War begins a change which has kept deepening to this day. Increasing weight of the State through the funding and organizing of fundamental research (not to speak of its seizure by the Army); industrialization of the management and administration of science; hierarchizing, division of work, parcellization of tasks, and, in particular, partitioning of fields, separation of theory and experiment; these are the main features of science to-day, specially pronounced in the case of contemporary physics[6]. These socio-political phenomena have deep consequences on scientific activity at its most 'internal' level, although some persist in thinking of it as neutral and pure. The division of scientific work, with the ensuing separation between tasks of (1) fundamental research (study of new concepts and phenomena), (2) 'fundamental development' (exploration and exploitation of the theoretical and experimental domains opened by fundamental research), and (3) teaching (in the broad sense, that is, spreading of scientific knowledge, including popularization) for instance, has impeded the recasting processes of modern physics. For it is usually through development and teaching that new theories are faced with practices which may first dissolve their archaic attle and then restructure them on a specific basis – under the condition, however, that these practices may act through a suitable theoretical feedback. The separation of the various scientific activities hinders the dialectics of such a process. While it was natural and implicit in the former mode of scientific research, recasting today can be but a specific activity, explicit and determined; it cannot escape from the very division of tasks that it criticizes. To state this contradiction rather than to ignore it, to use it as a tool rather than to be victim of it, such is my intention here.

The second feature of the particular conditions surrounding the springing up of quantum physics is its philosophical context. It is not uncommon that during periods of breakthrough in the history of science, philosophy comes to play an important role[4]; the criticism of old concepts, the elaboration of new theories cannot proceed in a purely deductive way from experimental 'data'. Such or such philosophical trend can play a role as a motor – or brake. The founding fathers of quantum mechanics thus relied explicitly onto a philosophical point of view which, through its numerous variants, can be said to be a *positivist* one. The major and seminal role played in its time by this philosophical current may be understood easily. By rejecting the intrinsic *a priori* validity of any previous theoretical concept and subordinating it to empirical investigation, the holders of this point of view could get rid of an apparently compulsory reference to the concepts of classical physics, as these exhibited their limitations. An operationalist approach enabled them to use as much as possible of the available experimental results, in a 'phenomenological' way, as to-day high-energy theorists would say. Such a methodology is quite clear in the building by Heisenberg of its 'matrix mechanics' from the frequencies of atomic spectra [7]. In other words, by relying on a positivist philosophical standpoint as a fulcrum, physicists could break with the iron-collar of another philosophical domination, that of a narrow mechanicist rationalism, which had reigned for several decades. Indeed the first attempts to a working positivism in physics, the ideas of Mach for instance, or of Ostwald, had been concluded by a relative failure; atomism had largely overcome energetism, and the cartesian description of the physical world 'par figures et mouvements' appeared unchallenged by the end of the XIXth century. The 'crisis' of relativity was but a false alarm and, far from endangering the building of classical physics, Einstein strengthened it by ensuring the consistency of electromagnetism with mechanics, within a reformed space-time. The quantum riddle was somewhat more serious . . . . Indeed, and offering a proof *a contrario* of the fecund importance of the positivist standpoint, those of the founders of quantum physics which stuck to the mechanicist rationalism of classical physics, such as Planck, Einstein, De Broglie, would not lead nor follow the major developments of quantum theory. They would not even accept them, or only to reconsider their opinion later on.

But – and this is my main point – the very same positivist current which had been so efficient to promote the breakthrough leading to the birth of quantum mechanics, rapidly turned into an obstacle, both epistemologically and pedagogically, for its recasting; the cornerstone had become a stumbling block. It will be my purpose in the following pages to try proving this statement. Let me already note here that the philosophical dogmas of the leading school were contradicted by their own supporters in their practice as physicists. For instance, it is Heisenberg himself who, after having emphasized the elimination of 'unobservables' elements from theoretical arguments as an epistemological golden rule of quantum physics,[3] some years later introduced the S-matrix notion with considerations upon the analyticity of its elements. How could one ever 'observe', or better 'measure' directly, an operator in an infinite-dimensional space, and analytic functions (that is, in particular, infinitely differentiable)? In fact, no physical *concept* can be directly measured or observed; as Feynman[4] writes sensibly (note that I do not refer here to philosophers' opinions):

It is not true that we can pursue science completely by using only those concepts which are directly subject to experiment. In quantum mechanics itself, there is a probability amplitude, there is a potential, and there are many constructs that we cannot measure directly (...) It is absolutely necessary to make constructs [8].

(the whole paragraph is worth reading).

To add one more argument yet for the necessity of the recasting that I advocate, I could propose a careful comparison of the ways a physicist thinks and talks according to whether on the one hand he *does* some quantum physics, with colleagues, dealing with his paper block or his apparatus, or, on the other hand, he *teaches* it, to students, in front of a blackboard. It is very rare that he uses, or simply mentions, in the first situation, the general philosophical statements that he steadily repeats in the second one. In other words, within this orthodoxy, as for most, there are many church-goers and few believers. Then, could say some people, the problem is not that serious. Is it really worthwhile fighting against ideas which are falling into abeyance and which are just paid lip service to? But it is precisely the most vulgar of the positivistic conceptions to consider philosophical and epistemological issues as deprived of interest, or of relevance, for the practicing of physics itself. Some praying mills are not as harmless as windmills, and it is not necessarily quixotic to tilt at them. Without any more preliminary justification, let me now try to sketch some directions for the recasting of quantum physics. For convenience, I will distinguish four types of problems, dealing respectively with the foundations, the description (terminology), the (so-called) interpretation and the (classical) approximation of quantum theory.

I. ON THE FOUNDATIONS OF QUANTUM THEORY

Down with the Correspondence Principle!

It is convenient to distinguish two different aspects in the foundations of quantum theories:

(1) *the Universal Framework*, that is, the set of general assertions, postulates and corollaries, which hold true for any quantum theory, irrespective of the particular physical situation to which it applies. As any theoretical structure of physics, it is not uniquely determined and obeys several formulations, with scopes of various extents. There exist old and narrow formalisms, such as the ones of the initial 'wave mechanics' or 'matrix mechanics', as well as modern and very general ones, such as the C*-algebra formalism. In between, and at the present stage, the Hilbert space formalism perhaps is the one with the wider use. In that formalism, a state of a physical system is represented by a vector (or, more precisely, by a ray) in a Hilbert space, the inner product of two such vectors yields a probability amplitude, the physical properties are represented by self-adjoint operators, etc. It is the collection of these rules, common to all quantum theories (within this formalism), which I call here the Universal Framework. In short, it is the part of the quantum theory which may be thought of as relying on the PRINCIPLE OF SUPERPOSITION as its cornerstone, or more generally, as corresponding to the linear structure of the theory. The generalization of the initial wave mechanics closely associated with classical wave theories through heuristic analogies, to the more general Hilbert space formalism is typical of a recasting process in the foundations of quantum theory, one among the few to have taken place. This aspect of the foundations has been considerably renewed in the last period by the work done on the so-called 'quantum logics'. These provide a new and deeper basis, although not completely stabilized yet, for the Universal Framework of quantum theory. Since these questions are dealt with at length in other contributions at this Colloquium, I will not insist any further, and will rather consider:

(2) *the specific structure* of particular quantum theories, describing restricted classes of physical systems, such as, for instance, 'non-relativistic' (Galilean) quantum mechanics of a particle (Schrödinger theory), many-body nonrelativistic theory, quantum electrodynamics, etc. For any such theory, the Universal Framework must be supplemented with specific assertions on the choice of the operators associated to the relevant physical properties, their algebraic relationships, the dynamical law of evolution for the system, etc. Chronologically speaking, the initial approach to these specific structures has been through the PRINCIPLE OF CORRESPONDENCE with the 'classical' theories. This is how nonrelativistic quantum

mechanics was built upon classical Hamiltonian mechanics, quantum electrodynamics upon Maxwell electromagnetism, etc. This approach was justified, indeed it was almost a necessary one in historical terms; some criterion of consistency with the old theoretical framework in effects is one of the strongest conditions to be imposed to any new, emerging theory, and can be followed as a trustworthy guide. But, despite its role as an Ariadne's clew, this umbilical chord should be cut some day, for the correspondence principle meets with several difficulties, theoretical and (epistemo)logical. For instance, either it is considered as a heuristic guide, such as it was used with fecundity by Bohr, but the scope and validity of which cannot be systematically assessed, or it meets with logical contradictions when given a precise theoretical formulation[10]. Much more serious is the fact that the correspondence principle gives us some knowledge of these quantum properties only which do possess a classical analog; specific quantum effects, vanishing in the classical limit, thus are outside of its scope. The quantized spin of 'elementary' quantum objects here is a conspicuous example. Another one is the concept of parity.[5] Finally, since we know a classical theory to have only approximate validity, in a much narrower domain that the corresponding quantum one, there seems to be some logical inconsistency in using the first one as foundations for the second. It is truly paradoxical to assert, as do Landau and Lifshitz, that:

A more general theory can usually be formulated in a logically complete manner, independently of a less general theory which forms a limiting case of it. (...) It is in principle impossible, however, to formulate the basic concepts of quantum mechanics without using classical mechanics. The fact that an electron has no definite path means that is has also, in itself, no other dynamical characteristics (*sic*). Etc. [11].

This vicious circle (see Figure 1) is directly linked to our lack of knowledge of the conditions of validity for the 'classical limit', about which some comments may be found in the last section.

Fortunately, the correspondence principle may be replaced, and advantageously so, by the use of the INVARIANCE PRINCIPLES. The Universal Framework of quantum theories, through the linearity properties, indeed endows the invariance groups of physical symmetries with a great importance[12]. It requires the existence of a unitary projective representation of the invariance group in the Hilbert space of any physical system with the relevant symmetry properties. A classification of these representations thus yields a

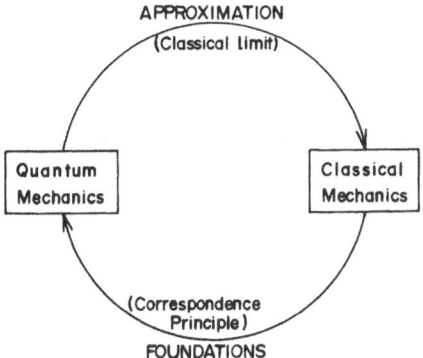

Fig. 1. The vicious circle.

classification of the possible quantum systems. Further, for Lie groups, through the SNAG theorem, any one-parameter subgroup is represented by a unitary subgroup derived through exponentiation from a self-adjoint operator, its 'infinitesimal generator'. These operators usually qualify for describing the most important physical properties, such as energy, momentum, etc. Other properties may then be found in the enveloping algebra of the Lie algebra by simple considerations of invariance (or, rather, group 'variance'). The position operator yields a simple and important example of this procedure [13]. From such a point of view, the concept of spin, for instance, appears in a very natural way, from a simple analysis of the space-time properties of an 'elementary' quantum object, whether it be in Einsteinian [14] or Galilean [15, 16] relativity. Besides deepening the foundations of the specific structure for quantum theories, the consideration of Invariance Principles, enable one to clear-up some old misunderstandings, as well as to shape open problems. The emphasis they deserve, it is true, is not a new discovery and has already been advocated in detail in the literature [16, 17], so that I will pass on to more controversial issues.

II. ON THE DESCRIPTION OF QUANTUM THEORY

In other words...

It is quite clear that in the actual practice of physics, no one can be content with the use of the sheer mathematical formalism, even though this formalism is a necessary and fundamental constituent of the considered theoretical domain [18]. A metalanguage is necessary as well, so that the names given to the mathematical objects and formal concepts of the theory enable its statements to fit in the general discourse. The choice of the terminology thus is a very delicate affair, with deep epistemological implications. If adequate, it may greatly help the understandability of the crucial points as it may hinder it in the contrary case. Now, the difficulty is that such a choice necessarily relies on abuses of languages or metaphors. Indeed one has to choose old names for new things: a completely invented name, created from scratch, would add nothing to the mathematical expression of a physical concept. One usually looks for a convenient name either in the ordinary language, or in the already specialized language of a previous theory. In the first case, we deal with a metaphorical analogy, for example when calling 'spin' the intrinsic angular momentum of a quantum particle. In the second case, it is an abuse of language to extend the name of a physical concept belonging to a certain theory, to a more or less homologous concept in another theory, for example when calling 'energy' the operator generating time-translations in quantum mechanics. Note that in that case, due to the Einstein-Planck relationship $E = \hbar\omega$, one could have used as well the classical term 'pulsation' for the same concept (within the numerically arbitrary constant $\hbar$). In fact, we deal here with a new, specifically quantum, concept, which may be given a name borrowed to classical mechanics, under the condition however that this abuse of language be explicitly recognized. In these two examples ('spin' and 'energy'), the choices of terminology may be considered as fortunate ones, giving in the first case a concrete picture (however approximate it is known to be), and in the second case a partially sound reference to a familiar limit theory with a large domain of validity. However, I contend here that, if for most such specific concepts of quantum theory the common terminology can be considered as adequate, the situation is much worse for the general concepts, belonging, one might say, to the Universal Framework. There, the weight is heavily felt of the philosophical prejudices which permeated the initial theoretical work in quantum physics. The recasting of quantum theory should incorporate at least a critique of the conventional terminology: to hope for its modification probably is unrealistic in the present state of sociological inertia of our profession.

Here is a partial list of the usual terms in quantum theory, with some of my reasons for rejecting them and possible alternate proposals (despite my skepticism on their implementability).

'*Observable*'. The word is a direct imprint in quantum physics of the positivism advocated by its founders. To call 'observable' any self-adjoint operator associated to a physical property of a system is a multiple nonsense. To start with, as I have already pointed out, no one will ever actually observe or measure such a highly complex mathematical being... As a matter of fact, the only physical quantities we do measure directly (with a few exceptions) are lengths: displacements of needles on measuring apparatus, tracks on photographs, etc. Already in classical physics, very few physical properties are 'observed'; think only of velocity for instance.[6] The very indirect link between physical measurements, or observations, on the one hand and the essential ideas in the analysis and understanding of the situation, precisely is the reason for there being a *theory* with *concepts*, that cannot be immediately (i.e. without mediation) expressed in empirical terms. Of course, the terminology ('observables') was introduced in an effort to overpass the limitations imposed by the formalism of classical mechanism. Since an analysis of the measurement process showed that one could not 'observe' simultaneously, for instance, the momentum and the position of a particle, one could forget about this classical requirement and concentrate upon the real... observables of the system, rather than imposing *a priori* theoretical notions. But the error has been in the ensuing confusion between the experimental description and the theoretical statements. Finally, we know so little, as I will emphasize in the following section, about an actual quantum theory of measurement, that very few such 'observations' of quantum properties can be theoretically analyzed. And, after all, we do care, experimentally speaking, for a handful only of such properties, those precisely which hold specific names (energy, momentum, spin, position, etc.). I would then rather use a general terminology such as '*physical properties*' in place of 'observables'. 'Dynamical variables' is acceptable also, though I find it a rather awkward expression, for instance when used to describe a... 'kinematical constant', such as energy for instance. Of course, it may be necessary to insist on the specific nature of these physical properties in a quantum theory, as opposed to classical ones. But, as I mentioned, 'observables' does not qualify for stressing the difference. Simply call them 'quantum properties', or 'q-properties', distinguished from classical ones or c-properties. Consider finally this terminological monster: 'commuting observables'; it associates a mathematical epithet with an empirical substantive – a true positivist chimera. Why not speak rather of 'compatible q-properties', or, on the formal side, of 'commuting operators'?

'*Observer*'. In most cases, this term simply is without any real theoretical function. It may be suppressed and, along with it, the whole sentence that contains it, without damage. In the few instances where it plays a role, it should be replaced, depending on the case, either by 'experimenter' (in general metaphysical discussions), or by 'measuring apparatus' (in epistemological or theoretical statements). These remarks will perhaps become clearer after the discussion below the quantum theory of measurement.

'*Uncertainties*'. Here is a case of mistaken borrowing to the vocabulary of experimental physics. When it was realized that, in quantum physics, a physical property of a system in general cannot be characterized by a sharp numerical value, the spreading of the possible values was assimilated to the experimental uncertainties on the classical physical properties. These c-properties indeed do have a sharp value in any physical state, but this value usually is only known up to some uncertainty defined by the experimental conditions (resolution of the apparatus, knowledge of auxiliary parameters). It should be realized to-day, however, that the essential formal difference between c-properties and q-properties is that the first ones 'are' numerical functions (at least in particle mechanics), while the second ones 'are' operators; that is to say, a q-property usually associates to a given physical state not a single numerical value but a whole spectrum. One does not deal with an empirical uncertainty, but with an intrinsic '*spread*', which could (and should) also be called '*spectrum width*', or '*dispersion*' or '*extension*' for instance. Of course, this is precisely what practicing physicists do: a Breit-Wigner curve is characterized by its width in any sensible laboratory talk; one only speaks of an energy uncertainty in classrooms. What is specific of quantum physics clearly is that *all* physical properties may have such a spreading. But already in classical physics some properties may not always be sharp. In classical wave theory, we know that a wave in general has a whole spectrum of frequencies, and, except for harmonic waves, not a single value. Nobody would think of calling the width of the pulsation spectrum an 'uncertainty' on the pulsation. How comes, then, that in quantum physics, the analogous energy width, $\Delta E = \hbar \Delta \omega$, becomes an 'uncertainty'? It is clearly seen here how the failure comes from not taking the quantum theory seriously enough, by keeping stuck to classical ideas irrelevant in the quantum domain. It has been argued frequently, by the Copenhagen school in particular, that the difference between classical and quantum physics is that the first one only is consistent with everyday intuition and common sense, so that our mental pictures and the words we use should necessarily be based on this classical realm. I can only answer by pointing to the situation a few centuries ago, when Aristotelian physics indeed was closer to common experience than the new, Galilean one; after all, arrows do not have an indefinite uniform motion, up and down are not physically equivalent, the Sun is observed to go round the Earth, etc. What happened since, is that, due to this new physics, our 'common' sense has been enriched and our 'intuition' has evolved. (See for instance note 6.) I contend that the same is true today and that half a century of practice in quantum physics should allow us to drop our classical prejudices.

Uncertainty principle', 'uncertainty relations'. Clearly there is no 'principle' here; the 'relations' between 'uncertainties' are but inequalities linking the dispersions in two non-compatible physical properties. These 'Heisenberg inequalities', as I propose to call them simply, are consequences of the true basic principles of quantum physics. As such they play a theoretically subordinate part. This is not to underestimate their importance. Quite on the contrary, I hold them for a major pedagogical and epistemological result[20]. Not only do they exhibit the essential difference between q-properties and c-properties, but they are a very effective heuristic tool. Precisely because they express the limits of validity of the classical concepts, they enable one to use classical expressions for approximating quantum derivations by imposing additional constraints which simulate the full quantum treatment. The Heisenberg inequalities, far from expressing intrinsic and final limitations to our physical knowledge, as many philosophers have commented upon them, quite on the contrary greatly help us to enrich and refine our understanding. A terminology relying on the idea of 'uncertainties' clearly cannot do full justice to this deep positive role, which is an additional reason to advocate a change. Finally, and to use once more a classical example for an argument *a fortiori*, has any one ever called the classical spectral inequality $\Delta k \cdot \Delta x \geqslant 1$ an 'uncertainty relation'? Why then, should the closely related quantum inequality $\Delta p \cdot \Delta x \geqslant \hbar$ receive this dubious privilege? This might be the place also to get rid of the so-called 'uncontrollable quantum perturbation of the observed system by the observer' (or the measuring apparatus), which are sometimes invoked as a source of the 'uncertainties'. It may be asserted simply that no such perturbations exist. Quantum theory does not imply a necessary (and unknown) change in the state of the system subjected to a measurement. In fact, most analyses of the measurement processes (see the next section), including the conventional and orthodox ones, use a simple model, going back to von Neumann, where the state of the measured system does not change.

'Complementarity'. We deal here with the typical example of a parasitical philosophical notion in physics. Not that it has been without utility: to physicists educated in classical mechanics, some general prescription was necessary to relieve them from the anxiety of not being able to apply any more classical ideas, such as the existence of simultaneously sharp numerical values for any two physical properties. When it became clear that quantum position and quantum momentum, for instance, decidedly did not fit into this scheme, they were interpreted as a pair of 'complementary' properties, the observation of one with arbitrary precision precluding that of the other. It is not a matter of observation of course, but rather a question of the fundamental nature of the quantum concepts. Only by insisting on their supposedly sharp numerical definition, do we need to introduce so vague a notion as the one of complementarity. Complementarity becomes a totally irrelevant idea for physics, as soon as one accepts the specifically quantum, i.e. qualitatively non-classical, nature of quantum theory. Also, the ever extended use to which complementarity was put to (by Bohr especially) should cast some doubt on the notion as a scientific one: from the complementarity between position and momentum, to that between particle and wave (see below), then to physics and biology, and even worse to society and individual, the ideological role of the idea becomes clear at last. Far from being an example of the philosophical impact of modern physics, a new way of thinking brought about by contemporary science, it is quite on the contrary a philosophical trojan horse inside physics and a witness of the real exploitation to which physics

has been subjected by some philosophical currents [4, 6]. The same type of critique will apply to my next target.

'*Wave-particle duality*'. Classical physics is built upon two key concepts, enabling it to describe all situations within its domain of validity: the concept of particle (discrete, localized) and the concept of wave (continuous, extended). But the objects the behaviour of which is described by quantum physics cannot be consistently analyzed in terms of these two concepts, although they share some of their characteristics. It was natural enough at the beginning of quantum physics to rely as far as possible on the known classical concepts, while using some criterion to avoid any situation where their contradictory properties would come into conflict. The point however is, as we should come to realize today, that the basic quantum objects are not either waves, or particles, but neither waves, nor particles. They must be described by some new concept, which, furthermore, turns out to be a unique one; several names have been proposed for such a concept, for instance 'wavicle' or 'quanton'.[7] In actual practice however, we still speak of 'particle', although we know well that they are not classical ones. Perhaps could we, at least in the beginnings of introductory courses, emphasize the point by writing 'partiqles'? It still remains to be said that the quantons have *something* to do with waves and particles ... But it is not that they appear either as waves, or as particles: as a matter of fact, most of the times they just appear for what they are, deserving a full quantum treatment, and not lending themselves to a classical wave or particle description. It is true, yet, that in some specific circumstances, there are valid wave or particle approximations (necessarily exclusive). The conditions for the validity of these approximations, although empirically more or less well-known, are not, to my knowledge, theoretically under control. It is an interesting problem, I think, to be solved, and one which should deepen our understanding of the quantum concepts as such (see also Section IV, below). Let me only note that bosons may obey either a wave, or a particle approximation; electromagnetic fields propagate as waves, and photons sometimes may be treated as classical particles. But fermions do not seem to have a classical wave description with any physical domain of approximate validity. And it is not clear whether the wave approximation for bosons does require or not a zero invariant mass (compare the cases of photons and pions). These last brief remarks are just intended to show how inadequate is the idea of a universal wave-particle duality, and how its very generality prevents one from dealing with concrete physical problems, namely studying the validity of *the classical wave and particle approximations*. To sum up, the 'wave-particle duality' is no more a correct way to analyze a quanton, than a 'rectangle-circle duality' would be to analyze a cylinder;[8] it is even worse, since the two partial aspects may hardly fit in the same picture with some consistency (Figure 2).

I will deal in the following section with the terminology of the so-called quantum theory of measurement: 'reduction of the wave-packet', 'perturbation of the observed system by the observer (or measuring apparatus)', 'indeterminism', etc., because of the need to discuss the whole question with some more details. I will not insist either on the very common incorrect use of some specific terms; for instance, 'wave-function', with the full weight of the classical metaphor it carries, should be exclusively restricted for denoting the state vector in the x-representation, $\langle x|\varphi\rangle = \varphi(x)$, and the same remark holds true for 'wave equations'. In more general situations, one should speak simply of the 'state vector', or the 'state', obeying 'dynamical' or 'evolution' equations. Many such examples can be found; while they do not need a thorough nor controversial analysis, I see no reason why we should tolerate a systematic sloppiness in our language. But let me end this section by calling into question, without any will to change it however, the 'quantum' label of quantum physics. It goes back, as we know well, to the discovery by Planck and Einstein of the discrete aspects of electromagnetism radiation, and was reinforced by the analysis of the quantized energy spectra of atoms, molecules, nuclei. The old saying "Natura saltus non fecit" seemed to be contradicted, and these discrete, quantal, aspects of the

Fig. 2. The wave-particle 'duality'. Although partial views of this figure may be interpreted as two-dimensional projections of three-dimensional bodies, the full figure is but a two-dimensional one, without such an interpretation.

new physics came to be thought of as its main characteristics. However, if it is true that in quantum physics, some continuous aspects of classical physics reveal their essential discontinuity, the converse is true as well. Instead of point-like particles, we deal with continuously extended quantons. Due to the tunnel effect, the transmission by a potential barrier which classically obeyed a simple yes-or-no law, is characterized by a coefficient with a value in the continuous range between 0 and 1. The quantum world is not 'more discontinuous' than the classical one.[9] Instead, while the classical world could neatly enough be divided into a continuous part (waves) and a discrete one (particles), the quantum world is a single one, where this opposition is a rather irrelevant one.[10]

III. ON THE 'INTERPRETATION' OF QUANTUM THEORY

The late hatching of a Columbus' egg

Most philosophical exegeses, commentaries and discussions about quantum theory and its interpretation up to now have been centered on the so-called 'measurement problem'. Stripped down to its essentials, the problem comes from the apparent contradiction between the two kinds of time-evolution followed by a quantum system:

(1) when isolated, a quantum system is described by a state-vector driven in the Hilbert state space by a linear unitary evolution operator: $\Psi(t) = U(t, t_0)\Psi(t_0)$. This operator U in turn is linked to the Hamiltonian which acts as the time-evolution generator. Such a behaviour, closely related to the validity of the superposition principle, is continuous and perfectly deterministic.

(2) when subjected to a measurement, however, the state vector is said to 'collapse' onto one of the eigenvectors of the operator associated to the physical property under measurement, and the measured value of the property is given by the corresponding eigenvalue. This so-called 'reduction of the wave-packet' is not a linear process in the state space of the combined system consisting of the measured system and the measuring apparatus (or ... 'observer') and cannot follow a deterministic unitary evolution of the preceding type. It is at this stage that the alleged 'indeterminism' of quantum theory enters: the projection of the state-vector (as I will say instead of 'reduction') onto one of the eigenvectors obeys a stochastic process,

with probabilities given by the squared modulus of the inner product between the initial state vector and the final eigenvector.

There is no need to stress that this probabilistic rule up to now has been supported by all the available evidence. It thus seems as if quantum theory was self-contradictory, since the behaviour it requires for isolated systems (type 1) could not apply to such a system when composed of a measured subsystem and a measuring one, for it has to obey a type 2 – evolution. Several solutions have been proposed out of this dilemma.

The conceptually simplest ones, apparently the most radical, but – to me, at least – in fact the most conservative, do not question the existence of the conflict, and explain it by a fundamental incompleteness of quantum theory. One may then look for a fundamental change, and investigate 'deeper' theories. Such is the goal of the various 'hidden variables' theories. Their strong classical flavour nevertheless makes it hard to believe that the difficulties of quantum physics might be solved in such a backwards way. However, there is no need anymore for lengthy philosophical discussions on this point; due to the work of Bell and others [24, 25], we now know that there are experimentally checkable differences between the predictions of quantum theory on the one hand, and hidden-variables theories on the other hand, unless these exhibit rather weird nonlocal features, which would plague them with conceptual problems even worse than the ones they are supposed to solve. Other possibilities derive from accepting quantum theory, but supplementing it with various external devices which would explain, through some more or less natural physical mechanism, the projection of the state vector: one may invoke specific 'physical' (!) laws obeyed by the mind of the living observer, as Wigner proposes [26], or, more soberly, macroscopic ergodicity [27], or still, gravitational fluctuations. But most of these attempts suffer from their rather 'ad hoc' character. Indeed it is difficult, if the measurement process is considered as an interaction process between a (measured) system and an apparatus (or even an observer), to put such processes entirely apart from all other physical interaction processes, and to understand how they could obey specific laws without there being testable consequences outside of measurement theory as such.

The dominant conception, at least, is a fully consistent one. The 'Copenhagen school' answer, or rather the way I understand it (for,

as in any orthodox church, the fundamental dogmas may be interpreted in thousands of ways), consists in eluding the physical problem
by giving it a philosophical solution. The state-vector receives a
purely subjective interpretation, as a mere recording of the known
informations on the system. Any new data, such as given by a
measurement, then obviously change this catalogue. This change is
not ruled by the laws of physics proper, it is a truly meta-physical
process. This positivist standpoint is free from contradictions, especially since, as von Neumann showed, the same results obtain
whether the projection is supposed to take place during the direct
measurement (of the system by the apparatus), or a following one
(recording of the apparatus state by another apparatus), or the final
observation (by a 'conscious' experimenter). I only wish to stress that
one cannot speak here of a 'measurement theory' since measurement
is precisely put apart of the physical processes to which quantum
theory applies. One should rather consider the projection rule as a
supplementary postulate of the theory, of an empirical nature, and
which can be shown to be consistent with the rest of the theoretical
structure, provided we accept a particular philosophical interpretation. It has been repeatedly emphasized by the founding fathers
of the theory that this interpretation implies rather drastic consequences for our world-view. For instance, because of the subjective
interpretation of the state vector, no objective properties can be attributed to quantum systems as such. In other words, no quantum ontology
is possible (see above the quotation by Landau and Lifshitz, p. 172).
My previous proposals for changes in the terminology ('physical properties' instead of 'observables', for example) would then meet with
a strong reluctance from the custodians of the orthodoxy; indeed, these
changes need a consistent re-interpretation of the 'measurement
problem', which I sketch below. Another inescapable feature of the
Copenhagen interpretation is its dualism: there must exist two separate physical worlds, a quantum one and a classical one. All
measuring apparatus and observers necessary follow the laws of
classical physics, and the theoretical predictions as well as the experimental results must be formulated in classical terms, as Bohr
specially pointed out. This is not a question of convenience due to the
macroscopic nature of most experimental devices which would imply
an approximately classical behaviour (see below a discussion of this
'approximation'). Rather it is a question of principle; there is no fixed

place for the classical quantum borderline and its location may be
arbitrarily moved provided it separates the measured object from the
ultimate observing device. By reading the original works of the old
masters, one cannot but admire the consistency and depth of their
views. It must be said that much of their thoroughness was gradually
lost by the following generations and that the customary statements
of this epistemology, in most textbooks for instance, usually fail, by
and large, to reach the original standards of rigour, clarity and
coherence. I claim here that this philosophical decay does not have its
only cause in the exceptional genius of the great masters, as compared to our present average level of understanding.

Instead, I would argue that we do not need any more to rely upon
these philosophical principles in order to further our work in quantum
physics. As I have already pointed out, the main problem of the first
generation of quantum physicists was to get rid of the epistemological
prejudices linked to classical physics, while at the same time relying
on the approximately valid aspects of the very same classical physics
as far as possible in the quantum domain.[11] The work of Bohr himself
is a splendid illustration of this point; for most of its great physical
contributions, he never used the fully developed and consistent
quantum formalism, but he worked out with the utmost cleverness
and an admirable insight semi-classical approximations, for instance,
in building derivations based on the correspondence principle. The
prescriptions of the Copenhagen School thus played a seminal role by
ensuring the necessary philosophical security to the first explorers of
the quantum domain, keeping them from falling back into classical
preconceptions as well as from asking premature questions in quantum theory.

Things have changed today, however, and – this is my leitmotiv,
indeed – we should try to draw the lesson of half a century of
quantum practice. For if most physicists only pay lip service to the
orthodox dogmas, it is that in fact they hold opposite beliefs, although
implicitly only. Most of us, in our daily laboratory work, do act as if
quantum systems in fact had objective existence and properties, as if
quantum physics was universally valid and classical physics only a
convenient approximation. For, raised in a quantum context (many of
us know much more of quantum than of classical physics), we do not
need to fight all day long against classical prejudices. Our physicists'
common sense, so to speak, is no longer contradictory with quantum

theory, the apparent 'paradoxes' of which do not trouble us any more. In other words, besides the dominant explicit neo-positivist interpretation of quantum physics, there is a no less widely shared implicit realist point of view. The recasting which I advocate here should consist in expliciting, strengthening and developing this point of view – in other words, transforming the general silent indifference regarding the orthodox position into a voiced difference.

A decisive step into that direction was accomplished by Everett[28], more than fifteen years ago, with the efficient support of Wheeler[29], later followed by several authors[30]. His solution to the vexed question of the state-vector projection in a measurement process is a very simple one indeed, namely that this projection does *not* occur . . . . The projection postulate, he showed, is not needed to obtain the usual results of quantum theory. I will sketch the idea on the usual simple example. Let S be a quantum system with two basis states φ_+ and φ_- (spin up and down, to follow the tradition). Let A be a measuring apparatus with initial state Φ_0; upon interacting with S, A goes into a final state Φ_+ (resp. Φ_-), if S is in the state φ_+ (resp. φ_-). Φ_+ and Φ_- may be thought of as macroscopic pointer states, indicating at the end of the measurement the initial state of the system S. In other words, the evolution operator U of the combined interacting system S & A, is defined by:

$$U(\varphi_\pm \otimes \Phi_0) = \varphi_\pm \otimes \Phi_\pm.$$

Consider now an arbitrary state of S, that is, a linear superposition, $\varphi = c_+\varphi_+ + c_-\varphi_-$. The combined system S $\oplus$ A, if in the initial state $\varphi \otimes \Phi_0$, will end in the final state:

$$\Psi = U(\varphi \otimes \Phi_0) = c_+\varphi_+ \otimes \Phi_+ + c_-\varphi_- \otimes \Phi_-,$$

according to the linearity of U ('type 1' evolution process). The projection postulate then asserts that the very act of measurement (or observation) somehow will cut off one of the two components of this state, and leave the combined system in one of the states $\varphi_+ \otimes \Phi_+$ or $\varphi_- \otimes \Phi_-$, with respective probabilities $|c_+|^2$ and $|c_-|^2$ ('type 2' process). Now, Everett points out, independently of this projection postulate, if A is to be a good measuring apparatus for the 'spin' of S, its pointer states Φ_+ and Φ_- certainly must be orthogonal, in order that the two possible states φ_+ and φ_- may be discriminated without ambiguity. If such is the case, it is well known, in the most orthodox tradition, that, *as far as the subsystem* S *is concerned*, the pure state Ψ is completely equivalent to the density matrix:

$$\rho = |c_+|^2 p_+ + |c_-|^2 p_-,$$

where $p_\pm$ are the projectors onto the states $\varphi_\pm$. It is now a purely subjective choice to interpret the state of S within the compound system $S \oplus A$ with state vector Ψ as being rather described either by φ_+ with a probability $|c_+|^2$ or by φ_- with a probability $|c_-|^2$. This interpretation, which is the one associated with the projection postulate, gives exactly the same theoretical results *for* S (for instance, the average values of any physical property) than the plain use of Ψ itself. Everett thus simply denies the need for the projection postulate, and gives a solution to the difficulties of the quantum theory of measurement which is really in the spirit of the Columbus' egg problem. One may now see the epistemological root of the projection postulate; it lies in the difficulty of fully accepting the superposition principle. Indeed, the intrinsic linearity of quantum theory, cannot be interpreted in classical terms. The quantum 'plus' which relates two superposed states, cannot be thought of as a classical 'or'. If one wants to fall back on this classical disjunction, then a supplementary assumption is necessary, extraneous to quantum theory as such. This precisely is the role played by the projection postulate, which allows a quasi-classical interpretation of the measurement process as yielding either such a result, *or* that one. Another way of saying this, is that, only by using the projection postulate, can we attribute a definite state to the system S after the measurement. The rejection of the postulate does not allow such a characterization, and we must deal with the non-separable state vector describing the compound system $S \oplus A$. It is precisely this specific quantum non-separability, upon which Everett rightly insists [28], which the projection postulate tries to bypass.

But how can all this be reconciled with our daily experience? After all we see either a down spot, or an up one in a Stern-Gerlach apparatus! The answer is simple – just treat an observer O as an apparatus; let χ_0 be its initial state and $\chi_\pm$ the final ones corresponding to the observation of the apparatus states $\Phi_\pm$. The combined system $S \oplus A \oplus O$ will now go from the initial state $(c_+\varphi_+ + c_-\varphi_-) \otimes \Phi_0 \otimes \chi_0$ to the final one $c_+\varphi_+ \otimes \Phi_+ \otimes \chi_+ + c_-\varphi_- \otimes \Phi_- \otimes \chi_-$. As long as χ_+ and χ_- are orthogonal, as they should be if the observation is to be a reliable one, no interferences can take place, that is, no mixing of the 'consciousness states' χ_+ and χ_-, which remain disjoint, each one being correlated to the correct system states φ_+ and φ_-. The very linearity of the evolution process entails the consistency of this scheme as Everett shows by giving examples of multiple measurements on a given system, as well as chains of measurements, each successive apparatus measuring the state of the preceding one.

Of course Everett's suggestion has far-reaching consequences, which he carries consistently and which are exactly opposite to those of the Copenhagen interpretation. There is now but a unique world, a purely quantum one. It is described by a universal state-vector (the title of Everett's big paper [28] is the Theory of the Universal Wave-Function'), which evolves according a single unitary deterministic process ('type 1'). A measurement, in this scheme, is but a specific type of interaction, the effect of which is to produce 'correlated' final states, namely, states of the form $\Sigma_k c_k \varphi_k \otimes \Phi_k$, where $\{\varphi_k\}$ and $\{\Phi_k\}$ are basis of the two subsystems, instead of the most general state $\Sigma_k \Sigma_l \gamma_{kl} \varphi_k \otimes \Phi_l$. Everett's analysis also sheds new lights on the probabilistic interpretation of the $|c_k|^2$, although it does not go as far as to have "the formalism dictate its own interpretation", as somewhat too enthusiastic supporters would make us believe [30] .... I will not comment on that point however, and would rather insist on what I believe to be a more serious misunderstanding of Everett's thesis by many of his followers. Once more, under a question of terminology lies a deep conceptual problem. The above interpretation in effect has been called by several people, especially De Witt, one of his main propagandists, the "many-worlds (or many-universes) interpretation of quantum theory" [30]. The rejection of the postulate projection leaves us with the 'universal' state vector. Since, with each successive measurement, this state-vector 'splits' into a superposition of several 'branches', it is said to describe 'many universes', one for each of these branches. Where the Copenhagen interpretation would arbitrarily choose 'one world' by cutting off all 'branches' of the state-vector except one (presumably the one we think we sit upon), one should accept the simultaneous existence of the 'many worlds' corresponding to all possible outcomes of the measurement. Now, my criticism here is exactly symmetrical of the one I directed again the orthodox position: the 'many worlds' idea again is a left-over of classical conceptions. The coexisting branches here, as the unique surviving one in the Copenhagen point of view, can only be related to 'worlds' described by classical physics. The difference is that, instead of interpreting the quantum 'plus' as a classical 'or', De Witt *et al.* interpret it as a classical 'and'. To me, the deep meaning of Everett's ideas is not the coexistence of many worlds, but on the contrary, the existence of a single *quantum* one. The main drawback of the 'many-worlds' terminology is that it leads one to ask the question of 'what branch we are on', since it certainly looks as if our consciousness definitely belonged to only one world at a time. But this question only makes sense from a classical point of view, once more. It becomes entirely irrelevant as soon as one commits oneself to a consistent quantum view, exactly as the question of the existence of the ether was deprived of meaning, rather than answered, by a consistent interpretation of relativity theory. In the words of Everett:

Arguments that the world picture presented by this theory is contradicted by experience, because we are unaware if any branching process, are like the criticism of the Copernican theory that the mobility of the earth as a real physical fact is incompatible with the common sense interpretation of nature because we feel no such motion. In both cases, the argument fails when it is shown that the theory itself predicts that our experience will be what in fact it was[28].

Of course, the very same analysis which shows that the projection postulate is unnecessary as a fundamental part of quantum theory, also shows that it is a convenient recipe in practical work. It allows one to deal with states characterizing the considered system alone, instead of the global state of the system – and – apparatus, not to say of the whole universe. Rather than a separate postulate, we should view it as a theorem, and a most useful one, the importance of which I do not intend to minimize. In other terms, a consistent treatment of quantum theory does not require the projection postulate, but everything works 'as if' it did hold.[12] Perhaps *not* everything, after all, since one of the reasons why people like Wheeler and De Witt support the Everett interpretation, is their belief that it may allow a conceptual welding of quantum theory with general relativity which seems difficult within the conventional treatment; this is another problem, upon which I do not want to comment here.

Yet this cannot be the end of the story. Indeed, Everett's reinterpretation breaks an important 'epistemological obstacle' (according to Bachelard's expression) on the way to a better quantum theoretical understanding. But it opens for us now the task of building concrete analyses of quantum measurements. As we have seen, the Copenhagen interpretation cannot but consider this question as a metaphysical one. On the contrary, from the new point of view, as I have already stressed, a measurement is a specific type of interaction process between two physical systems, which is to be fully described within quantum theory. It is to be proved, in such cases, that the system called 'apparatus' and its interacting with the 'measured' system, indeed possesses the characteristics necessary to its performance as a measurement device.

It is not sufficient, in that respect, that particular macroscopic pointer positions be described by orthogonal states of the apparatus, as the too sketchy analysis above might lead one to conclude. In fact, all non-diagonal matrix elements should vanish for every operator describing a reasonable physical property of the apparatus which could serve as a pointer for the considered measurement. In particular, one would like to understand, from that point of view, the role of the macroscopic nature of the measuring apparatus, which – contrarily to the Copenhagen orthodoxy – we do not consider as obeying classical mechanics. The challenge has been successfully met by Hepp who, for the first time, gave specific models of measurement processes, completely analyzed in quantum theoretical terms [31]. He showed, in his most realistic example, the so-called Coleman-Hepp model,[13] that the necessary orthogonality of the pointer states of the measuring apparatus for an adequate class of its macroscopic physical properties, results from a superselection rule in the relevant state space, obtained in the double limit where (i) the apparatus becomes infinitely extended (with an infinite number of particles), (ii) the interaction time (that is, the duration of the measurement) becomes infinite as well.[14] Of course, both these conditions never are met rigorously in practice.[15] However, Hepp could estimate the corrections due to the finite size and time of the measurement process; they are quite negligible. Hepp's analyses may and should be refined and extended to more realistic situations – as well as to simpler ones, perhaps, for educational purposes. But we know now that theoretical analyses of measurement processes can be developed in fully quantum terms – rather than a general theory of measurement, for which there is no place as a separated entity, if a measurement is but a specific type of physical interaction.

IV. APPROXIMATIONS OF QUANTUM THEORY

Back to Classical Physics

My main theme in the preceding sections has been the assertion that, to the present day, much of quantum physics foundations, terminology and interpretation, unduly relies on classical physics. I have also tried to explain the reasons for such a situation. As long as classical physics is used as a starting point towards quantum physics, their relationship hardly can be analyzed but in abstract philosophical terms, as in the Copenhagen view. On the contrary, as soon as quantum physics may stand on its own, its connection with classical physics may be subjected to theoretical analyses, rather than to meta-theoretical ones. And, indeed, it is a vast domain to investigate, in which deep and important physical problems too long have been obscured by epistemological prejudices. Surprising as it may seem, we do not have today a serious understanding of the classical approximation to quantum mechanics. There are, of course, formal derivations of the mathematical structure of classical mechanics from the one of quantum mechanics. They generally consist in studying mathematical limit processes in which Planck's constant vanishes. It is apparent that such processes are purely formal and, at most, give us a proof of the theoretical possibility that classical mechanics be a valid approximation to quantum mechanics.[16] But they tell us nothing about its physical conditions of validity. Besides the philosophical veto for such investigations as expressed by the conventional view of quantum physics, there may be another cause to this gap. It is commonly, although perhaps implicitly, thought that the problem is a simple one and that the classical/quantum dichotomy merely corresponds to the macroscopic/microscopic one. Once more, this idea reflects a past historical situation. For a long time, indeed, all known macroscopic systems could be analyzed by classical physics, while quantum physics was restricted to atoms and molecules, nuclei and fundamental particles. We know today, however, as a result of our long experimental and theoretical work in quantum physics, how to observe macroscopic quantum effects, in well-specified conditions. Physical systems such as lasers,. superconductors and superfluids, indeed exhibit clear quantum effects on a macroscopic scale.[17] Thus the large size (or, rather, number of particles) cannot be a sufficient condition for the validity of classical concepts. Neither is it a sufficient one, since classical (or semi-classical) approximations, to-day find an extended use, in fundamental particle physics for instance.

In fact, besides these spectacular but rather particular quantum effects, quantum physics plays an all-pervasive role in our everyday macroscopic world. For, not only is classical physics unable to ensure the stability of an isolated atom, a difficulty which was one of the sources of quantum physics, but it cannot explain the stability of their grouping into ordinary bodies, such as crystals for instance. In that respect, the relationship between quantum and classical (non-quantum) theory, is a much more complicated one than the relationship between 'relativistic' (Einsteinian) and classical (Galilean) physics. The Galilean theory of relativity has a wide scope of approximate validity, even extending far enough to include surprisingly many electromagnetic phenomena[36]. Above all, it provides a consistent (although erroneous) view of the world, to be contradicted only by rather elaborate experimental tests. On the contrary, classical mechanics loses its inner consistency as soon as it hits upon the atomic hypothesis; the extended bodies of our common experience, in classical terms, can only be thought of as perfectly homogeneous and continuous lumps of matter. Since the atomic hypothesis itself is grounded in well-known and very elementary chemistry and thermodynamics, it is seen that classical mechanics is contradictory with other parts of the classical picture of the universe. After all, its incapacity to provide an explanation of the black-body radiation, as a macroscopic failure, probably was a much more serious cause of concern than the puzzles associated to atomic spectra or the photoelectric effect.

It may be surprising, then, that the stability of ordinary matter has stood for so long before being proven on the basis of quantum theory. Only in the recent years has the problem been solved through the efforts of Dyson and Lenard[37], and Lieb and Lebovitz[38]. The first of these authors were able to show that in a system consisting of massive charged particles interacting via Coulomb forces, the binding energy per particle is bounded independently of the number of particles (saturation of forces), under the condition that at least the particles with one sign of their electric charges belong to a finite number of species of fermions. In other words, it is the Pauli principle ruling the electrons which ensures the stability of the world.

As a counter-example, Dyson also has shown that boson systems interacting via Coulomb forces are *not* saturated[39]. The specific quantum nature of the Pauli principle thus is a proof of the need for a quantum explanation of the most fundamental aspects of the physical world, namely its consisting of separate pieces of matter with roughly constant density. Pursuing this work, Lieb and Lebovitz were able to prove rigorously the existence of thermodynamic limits for the physical properties of interest in such Coulomb interacting bodies[38]. I hold these results for some of the most important ones in theoretical physics during the past years: they can be said to provide a real and deep explanation of very general physical phenomena, right from first principles. It is ironical enough that they do not use any recent knowledge, neither of empirical data, nor of mathematical techniques, and 'could' have been established a long time ago, were it not for epistemological obstacles. True, the analysis of Dyson and Lenard is a monument of subtlety, proceeding through a very long chain of clever inequalities. It is highly desirable that a new, shorter, proof be given to provide an easier access to the result and to bring down the estimate on the bound of the energy per particle to a more plausible value; due to the cumulative multiplying of the successive estimates, it is actually some 10^{14} times higher than the empirical value.[18] To still emphasize the highly non-trivial nature of these analyses, let it be said that the Coulomb potential precisely is a critical one: for potentials decreasing faster than r^{-1} at infinity and slower at the origin, saturation may be proved much more easily[41]. Or still, for purely attractive forces, such as those responsible for gravitation, saturation is trivially shown not to hold[42],[19] so that it is the delicate balance of attractions and repulsions in Coulomb systems which endow them with their very special properties. One may see here how, as I asserted earlier, the 'classical' behaviour does not result in a simple and universal way from some formal approximation to quantum theory, but requires, on the contrary, a thorough analysis of the specific physical situation.

One could also quote here other studies of the behaviour exhibited by various specific models of quantum systems in the macroscopic limit. Simple models of the collective and cooperative interactions of radiation and matter (such as based on the 'Dicke Hamiltonian'), have led to a better understanding of quantum optics and laser physics [44];[20] they provide an active and fruitful field of investigation.

The above-mentioned works deal with the possibly classical behaviour of macroscopic bodies for various specific physical situations. A more general approach to 'the connection between macro-
physics and microphysics' has been proposed by Fröhlich[47].
Starting right from the microscopic Schrödinger equation obeyed by
the density matrix of an N-body quantum systems, he studies how
various approximations, depending on the concrete situation, may lead
to macroscopic physical laws. In his own words:

> The method to use is to formulate the relevant macroconcepts, say hydrodynamic
> velocity field, mass density, etc. in terms of microproperties and then to employ the
> exact microequations of motion (without attempting to solve them) for the derivation of
> dynamical laws between the macroconcepts, e.g. the equations of hydrodynamics. Such
> derivations of macroscopic equations nearly always require imposition of certain
> assumptions which specify the particular situation.

The macroscopic physical quantities in fact are related to reduced
density matrices, the linked equations of motion of which obey an
ascending hierarchy, to be suitably cut off, depending on the approximation used. His methods enable Fröhlich to derive classical
laws when valid, such as the Navier-Stokes equations for fluids, as
well as macroscopic-quantum approximations, such as are necessary
to understand superconductivity or superfluidity. Further applications
appear to be possible, leading to new results, rather than to the
recovery of old ones, for instance in biological systems[48]. Indeed,
in the apt words of Fröhlich:

> It might be thought that all interesting macroscopic properties had been found long ago and
> that the derivation of their dynamical laws would be a matter of time, but not of very great
> interest. In contrast, however, it will be noted that the concept of macroscopic wave
> functions which dominates the properties of superfluids and superconductors had been
> discovered in recent years only.(...) It is quite obvious that a very large number of
> undiscovered macroconcepts does exist in situations which are removed from thermal
> equilibrium. For otherwise one should be able to derive by systematic methods the
> properties of all machines made of metal, say, since one has been able to formulate the
> basic laws referring to the atoms of metals[47].

Since the starting point is a set of exact microscopic equations of
motion, not containing any statistical assumptions, an appropriate
treatment should finally permit the introduction of thermodynamic
quantities and yield all relations that hold between them. Such an
ambitious program, implying a new justification for statistical mechanics in general, has yet to be carried out; Fröhlich still has shown
that the expectation was fulfilled for very weak interactions. Finally,
it is fitting to conclude this too brief description of a major work by

quoting its last paragraph, in which Fröhlich, apparently unaware of Everett's work, comes to the position advocated in the present paper concerning the interpretation of quantum mechanics (see the preceding section).

... This article should not be closed without emphasizing the exclusive status of the density matrix Ω of "the whole world". In standard use of quantum mechanics, the interpretation of a state vector, or of the corresponding density matrix rests on the introduction of an observer who interferes with it. If Ω refers to the whole world, then no such observer can exist. Hence Ω develops causally with time, containing all possible quantum-mechanical possibilities.[21] (...) It would be fascinating to speculate on the consequences of an Ω attributed to "the whole world"[47].

CONCLUSION

An exercise: recasting quantum zipperdynamics

Rather than to close this paper by a trite summarizing of the precedent considerations, I prefer to leave it open-ended, by trusting further developments onto the readers. As a neat example, the practical importance of which cannot be denied, an urging task in recasting quantum theory could consist in the rewriting of Zipkin's theoretical zipperdynamics[49]. Since this fundamental work has too long been ignored, I think it useful to have it partly reprinted below. It will be seen how the author bravely deals with a problem in macroscopic quantum physics,[22] thus defying the orthodox tradition, while keeping attached to this very same tradition in his use of a worn-out terminology. No doubt many further progresses in such a crucial area might be achieved through a consistent recasting, such as the understanding of the (epistemological) obstacles which too often block zippers at mid-course.

Laboratoire de Physique Théorique, Université Paris VII

THEORETICAL ZIPPERDYNAMICS

HARRY J. ZIPKIN

Department of Unclear Phyzipics, The Weizipmann Inziptute

INTRODUCTION

The fundamental principles of zipper operation were never well understood before the discovery of the quantum theory[1]. Now that the role of quantum effects in zippers has been convincingly demonstrated[2], it can be concluded that the present state of our knowledge of zipper operation is approximately equal to zero. Note that because of the quantum nature of the problem, one cannot say that the present state of knowledge is *exactly* equal to zero. There exist certain typically quantum-mechanical zero-point fluctuations; thus our understanding of the zipper can vary from time to time. The root mean square average of our understanding, however, remains of the order of h.

ZIPPERBEWEGUNG

The problem which baffled all the classical investigators was that of 'zipper-bewegung'[3], or how a zipper moves from one position to the next. It was only after the principle of complementarity was applied by Niels Bohr[4], that the essentially quantum-theoretical nature of the problem was realized. Bohr showed that each zipper position represented a quantum state, and that the motion of the zipper from one position to the next was a quantum jump which could not be described in classical terms, and whose details could never be determined by experiment. The zipper just jumps from one state to the next, and it is meaningless to ask how it does this. One can only make statistical predictions of zipperbewegung.

The unobservability of zipperbewegung is due, as in most quantum-phenomena, to the impossibility of elimination of the interaction between the observer and the apparatus. This was seriously questioned by Einstein who, in a celebrated controversy with Bohr, proposed a series of experiments to observe zipperbewegung. Bohr was proved correct in all cases; in any attempt to examine a zipper carefully, the interaction with the observer was so strong that the zipper was completely incapacitated[5].

THE SEMI-INFINITE ZIPPER

A zipper is a quantum-mechanical system having a series of equally spaced levels or states. Although most zippers in actual use have only a finite number of states, the semi-infinite zipper is of considerable theoretical interest, since it is more easily treated theoretically than is the finite case. This was first done by Schroedzipper[6] who pointed out that the semi-infinite series of equally spaced levels was also found in the Harmonic Oscillator discovered by Talmi[7]. Schroedzipper transformed the zipper problem to the oscillator case by use of a Folded-Woodhouse Canonical Transformation. He was then able to calculate transition probabilities, level spacings, branching ratios, seniorities, juniorities, *etc.* Extensive tables of the associated Racah coefficients have recently been computed by Rose, Bead and Horn[8].

Numerous attempts to verify this theory by experiment have been undertaken, but all have been unsuccessful. The reason for the inevitability of such failure has been recently proved in the celebrated Weisgal-Eshkol theorem[9], which shows that the construction of a semi-infinite zipper requires a semi-infinite budget, and that this is out of the question even at the Weizipmann Inziptute.

Attempts to extend the treatment of the semi-infinite zipper to the finite case have all failed, since the difference between a finite and a semi-infinite zipper is infinite, and cannot be treated as a small perturbation. However, as in other cases, this has not prevented the publishing of a large number of papers giving perturbation results to the first order (no one publishes the higher order calculations since they all diverge). Following the success of M. G. Mayer[10] who added spin-orbit coupling to the harmonic oscillator, the same was tried for the zipper, but has failed completely. This illustrates the fundamental difference between zippers and nuclei and indicates that there is little hope for the exploitation of zipperic energy to produce useful power. There are, however, great hopes for the exploitation of zipperic energy to produce useless research.

REFERENZIPS

1 H. Quantum, 'A New Theory of Zipper Operation which is also incidentally applicable to such minor Problems as Black Body Radiation, Atomic Spectroscopy, Chemical Binding and Liquid Helium'. *ZIP* **7**, 432 (1922).
2 H. Eisenzip, 'The Uncertainty Principle in Zipper Operation', *Zipschrift für Phyzip* **2**, 54 (1923).
3 I. Newton, M. Faraday, C. Maxwell, L. Euler, L. Rayleigh, and J. W. Gibbs, 'Die Zipperbewegung' (unpublished).
4 N. Bohr, 'Lecture on Complementary Zippers', Geneva Conference, 'Zippers for Peace' (1924).
5 P. R. Zipsel and N. Bohm, Einstein Memorial Lecture, Haifa Technion (1956).
6 E. Schroedzipper, 'What is a Zipper', Dublin (1950).
7 E. Talmi, *Helv. Phys. Acta* **1**, 1 (1901).
8 M. E. Rose, A. Bead, and Sh. Horn (to be published).
9 M. Weisgal and L. Eshkol, 'Zippereconomics', Ann. Rept. Weizipmann Inziptute (1955).
10 Metro G. Mayer, 'Enrichment by the Monte Carlo Method: Rotational States with Magic Numbers', *Gamblionics* **3**, 56 (1956).

NOTES

[1] I take issue here with T. Kuhn's ideas on the history of science[3].
[2] The date was not that convenient, obviously, and, as thirty years later, science was drafted on the battlefield rather than celebrated in Colloquiums ....
[3] The abstract of his 1925's seminal paper reads. "The present paper seeks to establish a basis for theoretical quantum mechanics founded exclusively upon relationships between quantities which in principle are observable"[7]. Ironically enough, Heisenberg uses this argument in an entirely mistaken, although most fecund way, to exclude from theoretical considerations "unobservables in principle" quantities such as ... the position of an electron! The later development of quantum theory proved this property to be perfectly observable by itself.
[4] Let me seize this opportunity to stress the importance for the recasting of quantum physics of the two introductory textbooks by Feynman[8] and Wichmann[9]. They are the first ones to break on some decisive points with an antiquated tradition and to contain some bold, although often implicit, new points of view. I owe to them much of my personal understanding of quantum physics – which came much later than my learning it.
[5] Observe that parity cannot have a classical limit as a conventional mechanical property; for a system such as the hydrogen atom, its value for consecutive levels is alternatively $+1$ and -1, so that no well-behaved limit exists for large quantum numbers, when the levels crowd together.
[6] The birth and life in physics of the concept of (instantaneous) velocity offers a simple and convincing example of the recasting process. It can be said that this concept is the crucial point of the Galilean breakthrough which brought physics from a pre-scientific stage to the state of a true scientific theory. However, since Galileo could use but the Euclidean geometrical theory of proportions as a mathematical tool, it is no surprise that he had to struggle for many years to master the concept[19]. The development of mathematical analysis in the following century, and the rigorization of limiting processes, would later on endow the concept with a much more convenient formal expression. But the ultimate stage in this recasting process was only reached in the present century, when the theoretical concept was materialized, so to speak, in common solid apparatus, such as the speedometer which is to be found on hundreds of millions of cars. Thanks to this realization, any six-years old kid (well, at least in that small fraction of the humanity where cars are a usual commodity) does know, in empirical terms, that a speed of 60 m.p.h. does *not* mean that the car will run for 60 miles in an hour. The instantaneous nature of the velocity is visibly conveyed by the motions of the needle

under a quick acceleration or a brutal braking. A practical grasp of the concept thus builds on, at a collective level, easing the way for a later theoretical study. My contention is that a similar evolution is now taking place for quantum mechanical concepts, although, of course, on a socially much narrower scale.

[7] As advocated by M. Bunge, whose careful discussion of several issues contemplated here is very close in spirit to mine[21].

[8] I could also use another metaphor: the activities of several among our eminent colleagues might be analyzed by some people in terms of a 'scientific-military duality' according to whether they give (or rather sell) talks at Colloquiums such as this one, or advices to the Pentagon in the Jason division for instance[22]. On the contrary, I hold that these are but two aspects of a single and consistent sociopolitical situation, which, beyond the individual cases, is that of our whole professional community[6].

[9] A detailed study of the real peculiarities shown by quantum physics with respect to classical physics, with a healthy criticism of many commonly accepted ideas, has been worked out by M. Bunge and A. Kalnay[23].

[10] Conversely, one might as well call into question the name of 'classical physics' customarily given to pre-relativistic and pre-quantum physics. Have not 'relativity' theory and 'quantum' theory become 'classical' as well, after more than half-a-century of active development? Is it not true that most physicists to-day are much better educated in these sectors of so-called 'modern' physics, than in several important fields of 'classical' physics, such as hydrodynamics, for instance?

[11] Let me quote here the apt words of d'Espagnat for characterizing Bohr's views: "along with many satisfactory aspects, such a view has the well-known but nevertheless surprising feature of expressing the laws of the microworld by using approximate classical concepts referring essentially to *our* experience of the macroworld"[25]. I would only add that this experience of the macroworld was the only experience of the world available in Bohr's time, while we may rely today on a thorough experience of the microworld as well, which should entail the possibility of expressing its laws in a specific way.

[12] To pursue the analogy used in Everett's quotation above, where his interpretation of quantum theory is compared to the Copernican system, I would compare the common Copenhagen interpretation to the clever system devised by Tycho Brahé, who, by the way, was a Dane as well. His system was a compromise between the Ptolemaic system (here to be likened to classical physics), and the Copernican one; it had the earth fixed at the centre of the universe, with the sun circling around it and all other planets then circling around the sun. It is clear that this system is consistent with the more general Copernican one as it only supplements it with a choice of a particular privileged reference frame. The choice is unnecessary but convenient from the observer's viewpoint. The same exactly may be said for the projection postulate with respect to general quantum theory.

[13] Bell has given an elementary version of the model[33].

[14] Let me stress the need for the second condition (infinite duration of the measurement), perhaps more unexpected than the first one. Although models are possible where it is not required, they seem to be much too crude and physically irrelevant. Rather, a moment of reflexion will convince oneself that this condition indeed closely corresponds, as the first one, to the usual experimental situations. Also its importance comes from its contradicting the very general assumptions under which d'Espagnat has derived 'anti-quantum' Bell's type inequalities[25].

[15] It has been argued by Bell[33], against Hepp's point of view, that the limits (in size and time) necessary for the validity of the analysis are purely formal ones and that no 'rigorous' projection of the state vector actually occurs. Bell exhibited a physical

property of the measuring apparatus in the Coleman-Hepp's model for which the non-diagonal matrix elements do not vanish in the above limit. It is, Bell admits, a complicated object, with a strange time dependence, but its very existence, he maintains, prevents one from speaking about a 'wave-packet reduction'. Of course, I do agree with him, since I hold that there is *no* such 'reduction'! But what Bell holds for a drawback of Hepp's analysis to me is one of its assets, since it shows that not every macroscopic property of a given apparatus, but only a specific, though large, class of such properties may be used as efficient measuring pointers for a given property of the measured system. Indeed there are no universal measuring apparatus and experimenters usually stick to rather stable (rather than weirdly time-dependent) properties of their devices as reliable pointers!

[16] Hepp has given one of the most interesting analyses of this type, by studying the relationship between the limits $\hbar \to 0$ and $N \to \infty$ for quantum mechanical correlation functions[34]. He himself has emphasized that 'the classical limit is not unique' and that, even in the simple cases he considers, one may obtain the classical mechanics of N point particles as well as a classical field theory, depending on formal assumptions (see also the discussion of the so-called 'wave-particle duality' in Section II above).

[17] A nice example is given by the quantization of the vorticity in rotating superfluid helium. The circulation of the velocity vector around the vortices strings is quantized in units of value h/m, where m is the mass of the helium atom. Numerically, $h/m =$ 10^{-3} cm^2 s^{-1}, which means that this quantum effect takes place on a scale of a tenth of a millimeter in space and a tenth of a second in time. Indeed they may be observed almost with the naked eye ... [35].

[18] On the very day when I was finishing the present paper, such a proof appeared in print[40].

[19] Non-trivial results come out of a direct quantum study of macroscopic bodies consisting of gravitationally bound particles[43].

[20] Rigorous studies of the macroscopic thermodynamics (equilibrium and non-equilibrium) of such models have been achieved with exciting (!) results, such as the appearance of phase transitions (superradiance?)[45]. Unfortunately, it appears that these features were due to unphysical drastic simplifications in the original model[46].

[21] Let us note here that Fröhlich, in stressing, as he does, the quantum nature of the various 'possibilities', does not fall into the classical trap of the 'many-universes' terminology.

[22] One might also mention here the penetrating quantum-theoretical analysis of ordinary ghost phenomena by Wright, as a further proof of the importance of quantum effects in everyday life[50].

BIBLIOGRAPHY

[1] Planck, M., *Ann. Phys.* **4** (1901), 533.
   Einstein, A., *Ann. Phys.* **17** (1905), 132.
   Bohr, N., *Phil. Mag.* **6** (1913), 26.
   Heisenberg, W., *Zeits. Phys.* **33** (1925), 879.
   Born, M. and Jordan, P., *Zeits. Phys.* **34** (1925), 858.
   Dirac, P. A. M., *Proc. Roy. Soc. A*, **109** (1925), 642.
   Schrödinger, E., *Ann. Phys.* **79** (1926), 361; **79** (1926), 489; **80** (1926), 437; **81** (1926), 109.
[2] De Broglie, L., *Recherches sur la Théorie des Quanta*, thèse de Doctorat (Paris,

25.11.1924). Reprinted recently (Masson, Paris, 1963). English translation in *Selected Readings in Physics-Wave Mechanics*, G. Ludwig ed. (Pergamon Press, Oxford, 1968).

[3] Kuhn, T., *The Structure of Scientific Revolutions* (University of Chicago Press, Chicago, 1962).

[4] Althusser, L. *et al.*, *Cours de Philosophie pour Scientifiques* (Maspero, Paris, 1975); Fichant, M. and Pécheux, M., *Sur l'Histoire des Sciences* (Maspero, Paris, 1969).

[5] Maxwell, J. C., *Phil. Trans.* **155** (1865), 459.

[6] Lévy-Leblond, J. M., in *Ideology and the Natural Sciences*, H. and S. Rose eds. (Macmillan, London, 1976). See also (*Auto*)*Critique de la Science*, A. Jaubert and J. M. Lévy-Leblond eds. (Le Seuil, Paris, 1973).

[7] Heisenberg, W., paper quoted in Ref. [1], translated in English in *Sources of Quantum Mechanics*, B. L. Van der Waerden ed. (North-Holland, Amsterdam, 1967).

[8] Feynman, R. P., Leighton, R. B., and Sands, M., *The Feyman Lectures in Physics*, vol. 3 (Addison-Wesley, New York, 1965).

[9] Wichmann, E., 'Quantum Physics', *Berkeley Physics Course*, Vol. 4 (McGraw Hill, New York, 1971).

[10] Cohen, L., in *Contemporary Research in the Foundations and Philosophy of Quantum Theory*, C. A. Hooker ed., (D. Reidel, Dordrecht, 1973) pp. 71-76 and original references therein.

[11] Landau, L. D. and Lifshitz, E. M., *Quantum Mechanics* (Pergamon Press, Oxford, 1958).

[12] Wigner, E. P., *Proc. Am. Phil. Soc.* **93** (1949), 521; *Physics Today* **17** (1964), 34; *Symmetries and Reflections* (Indiana University Press, Bloomington 1967); *Bull. Am. Math. Soc.* **74** (1968), 793.

[13] Newton, T. D. and Wigner, E. P., *Rev. Mod. Phys.* **21** (1956), 400. Wightman, A. S., *Rev. Mod. Phys.* **34** (1962), 845.

[14] Wigner, E. P., *Ann. Math.* **40** (1939), 149.

[15] Lévy-Leblond, J. M., in *Group Theory and its Applications*, vol. 2, E. Loebl ed. (Academic Press, New York, 1971) and further references therein.

[16] Lévy-Leblond, J. M., *Riv. Nuovo Cimento* **4** (1974), 99.

[17] Stein, H., in *Paradigms and Paradoxes*, R. G. Colodny ed. (University of Pittsburgh Press, Pittsburgh, 1972).

[18] Lévy-Leblond, J. M., 'Physique et Mathématique', *Encyclopaedia Universalis* (Paris, 1971).

[19] Ravetz, J. R., 'Galileo and the Mathematisation of Speed', in *La Mathématisation des Doctrines Informes*, G. Canguilhem ed. (Hermann, Paris 1972).

[20] Lévy-Leblond, J. M., *Bull. Soc. Fr. Phys.* **14** (*Encart Pédagogique* 1) 1973.

[21] Bunge, M., *Philosophy of Physics* (D. Reidel, Dordrecht, 1973).

[22] *Scientia* **107** (1972), 801; *Nature* **239** (1972), 182; *Physics Today* **25** (Oct. 1972), 62 and **26** (April 1973), 11; *Science* **179** (1973), 459 and **180** (1973), 446.

[23] Bunge, M. and Kalnay, A., *Int. J. Theor. Phys.* (in press).

[24] Bell, J. S. *Physics* **1** (1964), 195.

[25] D'Espagnat, B., *Phys. Rev.* **11** (1975), D 1424, and further references therein.

[26] Wigner, E. P., in *The Scientist Speculates*, I. J. Good ed. (Heinemann, London, 1962).

[27] Daneri, A., Loinger, A., and Prosperi, G. M., *Nucl. Phys.* **33** (1962), 297; *Nuovo Cimento* **44B** (1962), 119; Loinger, A., *Nucl. Phys. A* **108** (1968), 245.

[28] Everett, H., *Rev. Mod. Phys.* **29** (1957), 454, and original work, both reprinted in Ref. [30].

206   J.-M. LÉVY-LEBLOND

[29] Wheeler, J. A., *Rev. Mod. Phys.* **29** (1957), 463, reprinted in Ref. [30].
[30] Everett, H., Wheeler, J. A., De Witt, B. S., Cooper, L. N., Van Vechten, D., and Graham, N., *The Many-Worlds Interpretation of Quantum Mechanics*, B. S. De Witt and N. Graham eds. (Princeton University Press, Princeton, 1973).
[31] Hepp, K., *Helv. Phys. Acta* **45** (1972), 237. See also the discussion in Ref. [32].
[32] Bona, P. *Acta Phys. Slov.* **23** (1973), 149.
[33] Bell, J. S., *Helv. Phys. Acta* **48** (1975), 93.
[34] Hepp, K., *Commun. Math. Phys.* **35** (1974), 265.
[35] Steyert, W. A., Taylor, R. D., and Kitchens, T. A., *Phys. Rev. Lett.* **15** (1965), 546. Putterman, S. J., *Superfluid Hydrodynamics* (North-Holland, Amsterdam, 1974), Section 10.
[36] Le Bellac, M. and Lévy-Leblond, J. M., *Nuovo Cimento* **14B** (1973), 217.
[37] Dyson, F. J. and Lenard, A., *J. Math. Phys.* **8** (1967), 423; Lenard, A. and Dyson, F. J., *J. Math. Phys.* **9** (1968), 698.
[38] Lebowitz, J. L. and Lieb, E. H., *Phys. Rev. Lett.* **22** (1969), 631; Lieb, E. H. and Lebowitz, J. L., *Adv. Math.* **9** (1972), 316.
[39] Dyson, F. J., *J. Math. Phys.* **8** (1967), 1538.
[40] Lieb, E. H. and Thirring, W., *Phys. Rev. Lett.* **35** (1975), 687.
[41] Fisher, M. E. and Ruelle, D., *J. Math. Phys.* **7** (1966), 260.
[42] Lévy-Leblond, J. M., *J. Math. Phys.* **10** (1969), 806; *J. Physique* **30** (1969), C3-43; *Phys. Rev.* **1** (1970), D 1837.
[43] Hertel, P., Narnhofer, H., and Thirring, W., *Commun. Math. Phys.* **28** (1972), 159; Thirring, W., *Acta Phys. Austr.*, Suppl. XI (1973), 493.
[44] Allen, L. and Eberly, J. H., *Optical Resonance and Two-Level Atoms* (Wiley, New York, 1975), Chap. 7; Louisell, W. H., *Quantum Statistical Properties of Radiation and Applied Optics*, (Wiley, New York, 1973); Nussenzveig, H. M., *Introduction to Quantum Optics* (Gordon and Breach, New York 1973); Haake, F., in *Springer Tracts in Modern Physics*, edited by G. Höhler (Springer, Berlin, 1973), Vol. 66; Sargent, M., Scully, M. O., and Lamb, Jr., W. E., *Laser Physics* (Addison-Wesley, Reading, Mass., 1974); Agarwal, G. S., in *Springer Tracts in Modern Physics*, edited by G. Höhler (Springer, Berlin, 1974), Vol. 70.
[45] Hepp, K. and Lieb, E. H., *Ann. Phys.* **76** (1973), 360; *Helv. Phys. Acta* **46** (1973), 573.
[46] Rocca, F. *et al.* (to be published); Rzazewski, K., Wodkiewicz, K., and Zakowicz, W., *Phys. Rev. Lett.* **35** (1975), 432.
[47] Fröhlich, H., *Riv. Nuovo Cimento* **3** (1973), 490.
[48] Fröhlich, H., *Phys. Letters* **51A** (1975), 21, and additional references therein.
[49] Zipkin, H. J., *J. Irrepr. Results* **3** (1956), 6. Reprinted in *A Random Walk in Science*, R. L. Weber and E. Mendoza eds. (Institute of Physics, London, 1973), p. 86.
[50] Wright, D. A., *Worm Runner's Digest* **12** (1971), 95. Reprinted in *A Random Walk in Science*, R. L. Weber and E. Mendoza eds. (Institute of Physics, London, 1973), p. 110.

D. BOHM AND B. HILEY

ON THE INTUITIVE UNDERSTANDING
OF NON-LOCALITY AS IMPLIED BY
QUANTUM THEORY*

ABSTRACT. We bring out the fact that the essential new quality implied by the quantum theory is non-locality; i.e. that a system cannot be analyzed into parts whose basic properties do not depend on the state of the whole system. We do this in terms of the causal interpretation of the quantum theory, proposed by one of us (D. B.) in 1952, involving the introduction of the 'quantum potential', to explain the quantum properties of matter.

We show that this approach implies a new universal type of description, in which the standard or canonical form is always supersystem-system-subsystem. In the quantum theory, the relationships of the subsystems depend crucially on the system and supersystem in which they take part. This leads to the radically new notion of unbroken wholeness of the entire universe. Nevertheless, special contingent states of relative independence of the behaviour of the subsystems are possible, and this explains why the classical analysis into independent parts is a good approximation in certain contexts.

We illustrate these ideas in terms of the experiment of Einstein, Rosen and Podolsky, and also in terms of the properties of superfluid helium.

Finally, we discuss some of the implications of extending these notions to the relativity domain. In doing this, we indicate a novel concept of time, in terms of which relativity and quantum theory may be eventually brought together.

1. INTRODUCTION

It is generally acknowledged that the quantum theory has many strikingly novel features, including discreteness of energy and momentum, discrete jumps in quantum processes, wave-particle duality, barrier penetration, etc. However, there has been too little emphasis on what is, in our view, the most fundamentally different new feature of all; i.e. the intimate interconnection of different systems that are not in spatial contact. This has been especially clearly revealed through the, by now, well known experiment of Einstein, Podolsky and Rosen (EPR)[1, 2] but it is involved in an essential way in every manifestation of a many body system, as treated by Schrödinger's equation in a $3N$ dimensional configuration space.

Recently, interest in this question has been stimulated by the work

J. Leite Lopes and M. Paty (eds.), Quantum Mechanics, a Half Century Later, 207–225. All Rights Reserved

of Bell[3], who obtained precise mathematical criteria, distinguishing the experimental consequences of this feature of 'quantum interconnectedness of distant systems' from what is to be expected in 'classical type' hidden variable theories, in which each system is supposed to be localizable; i.e. to have basic qualities and properties that are not dependent in an essential way on its interconnections with distant systems.

Generally speaking, experiments inspired by this work have tended fairly strongly to confirm the existence of quantum interconnectedness[4], in the sense that they do not fit 'localizable' hidden variable theories of the type considered by Bell and those who follow along his lines. This work has been valuable in helping to clarify the whole question of quantum interconnectedness. However, it now presents us with the new challenge of understanding what is implied about the nature of space, time, matter, causal connection, etc.

If we are completely satisfied with one of the usual interpretations of quantum mechanics, then we may miss the full significance of this challenge. Thus, we may accept the rather common notion that quantum mechanics is nothing more than a calculus or a set of rules for predicting results that can be compared with experiment. Or else we may accept Bohr's principle of complementarity[5], implying that there is no way to describe or understand any underlying physical process that might connect successive events on a quantum jump (e.g. from one stationary state to another). Or we may adopt some of the theories involving a new quantum logic[6]. All of these approaches, although different in important aspects, have in common the conclusion that quantum mechanics can no longer be understood in terms of imaginative and intuitive concepts[7]. As a result, we are restricted, in one way or another, to dealing with the by now fairly well established fact of quantum interconnectedness in terms of abstract mathematical concepts. The special implications of these concepts cannot however be fully and adequately grasped, because at this level of mathematical abstraction, the property of 'quantum interconnectedness' does not appear to be very different from that of 'classical localizability' (the difference showing up mainly in rather complicated inequalities of the type derived by Bell[3] for certain kinds of experimental results).

Thus, when experiments confirm that there are no 'localizable' hidden variables, one's first reaction may tend to be of surprise, perhaps even of shock, because the notion of localizability has been so deeply incorporated into all our intuitive concepts of physics. However, as soon as we are content with non-intuitive and non-imaginative mathematical formulations, it seems that the surprise and shock suddenly vanish, because we are back to the familiar and reassuring domain of calculating experimental results with the aid of mathematical equations. Nevertheless, when one thinks of it, one sees that the implications of quantum interconnectedness remain as novel and surprising as ever. Thus, we have merely diverted our attention from the challenge of understanding what all this means.

If we had a consistent way of looking at the significance of quantum mechanics intuitively and imaginatively, this would help us to avoid allowing our attention to be thus diverted from the novel implications of quantum interconnectedness. Such a way of looking might also help to indicate new directions of theoretical development, involving new concepts and new ways of understanding the basic nature of matter, space, time, etc. In this connection, it might even be useful to consider theories that we do not regard as adequate on general grounds, or as definitive in any sense. For such theories may still give imaginative and intuitive insight in a situation in which there is at present no other way to obtain this. It is in this spirit that we are proceeding in the present paper. That is to say, we are not attempting here to make statements about what actually is the nature of reality, but rather we are merely looking at certain imaginative and intuitive concepts, to see what light they can shed on the new quality of quantum interconnectedness.

In particular, we are going to use as a point of departure, the so-called causal interpretation of quantum mechanics, developed by one of us[8, 9] in 1951. Although what we shall do here was to some extent implicit in the earlier work, it was not brought out adequately there. In addition, we feel that we have seen some further points, which enable us to lead on to new directions of development.

The most important of these points is that our work brings out in an intuitive way just how and why a quantum many-body system cannot properly be analyzed into independently existent parts, with fixed and determinate dynamical relationships between each of the parts. Rather, the 'parts' are seen to be in an immediate connection, in which their dynamical relationships depend, in an irreducible way, on the state of the whole system (and indeed on that of broader systems in which they are contained, extending ultimately and in principle to the entire universe). Thus, one is led to a new notion of *unbroken wholeness* which denies the classical idea of analyzability of the world into separately and independently existent parts. Through this, a novel direction is indicated for our general intuitive and imaginative thinking, which takes it beyond the limits imposed by classical concepts. And so it becomes possible to understand one and the same physical content *both* intuitively *and* mathematically, not only for classical laws, but also for quantum laws.

2. BRIEF REVIEW OF CAUSAL INTERPRETATION OF THE QUANTUM MECHANICS

In this section, we shall give a brief review of the causal interpretation of the quantum mechanics.

The essential feature of this interpretation [8] was the proposal that what are commonly regarded as the fundamental constituents of matter (e.g. electrons, protons, neutrons, etc.) are *both* waves *and* particles, in a certain kind of interaction (which we shall describe more fully in the course of this paper). We begin here with a discussion of the one-body system. Thus, we suppose, for example, that an individual electron is a particle with well-defined co-ordinates, x, which are functions of the time, t.

However, we postulate further that there is also a new kind of wave field, $\psi(x, t)$, that is always associated with this particle, and that is just as essential as is the particle for understanding what the 'electron' is and what it does.

To show how the wave field acts on the particle, we write the complex function, $\psi(x, t)$ in the form $\psi = Re^{iS/\hbar}$ where R and S are real. Then (as shown in Ref. [8]), Schrödinger's equation reduces to

(1) $$\frac{\partial P}{\partial t} + \text{div}\left(P\frac{\nabla S}{m}\right) = 0,$$

with

(2) $$P = R^2 = \psi^*\psi,$$

(3) $$\frac{\partial S}{\partial t} + \frac{(\nabla S)^2}{2m} + V + Q = 0,$$

where V is the classical potential and Q is a new 'quantum potential' given by

(4) $$Q = -\frac{\hbar^2}{2m}\frac{\nabla^2 R}{R}.$$

Evidently, Equation (1) may be taken as an expression of conservation of the usual probability density of particles $P = \psi^*\psi$ with a current given by $\mathbf{j} = P(\nabla S/m)$. So one can take

(5) $$\mathbf{v} = \frac{\nabla S}{m}$$

as the mean velocity of the particles. In the 1951 papers [8] this was taken as the actual velocity also, but in a later paper [9], it was suggested that the particle executes some sort of random movement (resembling Brownian motion) with an average velocity given by (5) and it was shown that the probability density, $P = \psi^*\psi$ is the 'steady state' distribution ultimately resulting from the random movements of the particle.

The essential new features of the quantum theory first show up in Equation (3). In the classical limit ($\hbar \to 0$) we may neglect the quantum potential (4) and then (3) reduces to the classical Hamilton-Jacobi equation, for a particle under the action of a classical potential V. So we are led to propose that more generally, the particle is also acted on by the 'quantum potential' Q, which of course depends on the new 'Schrödinger field' ψ, and which cannot be neglected in processes in which a single quantum of action, $\hbar$, is significant.

By way of illustrating how this theory works for the one-body system, we shall mention two essential aspects of the quantum theory here. Firstly, there is the well known phenomenon of interference of electrons (e.g. in a beam that has passed through several slits). The fact that no electrons arrive at points where the wave function is zero is explained simply by the infinite value of the quantum potential,

$$Q = -\frac{\hbar^2}{2m}\frac{\nabla^2 R}{R},$$

which repels particles and keeps them away from points at which $R = 0$. Secondly, let us consider the phenomenon of barrier penetration. As shown in detail in Ref. [8], the wave function for such a system is a time dependent packet. As a result, the quantum potential,

Q, fluctuates in such a way that occasionally it becomes negative enough to cancel the positive barrier potential, V, so that from time to time a particle may pass through before the quantum potential changes to a significantly less negative value.

In a rather similar way, all the essential features of the quantum mechanical one-body system were explained in Ref. [8]. Such an explanation was made possible merely by adding the concept of a wave field, ψ, to that of the particle, and by relating these two through the quantum potential, Q. Although some new features have thus been brought in relative to classical physics, there is still nothing foreign to the general classical conceptual structure, involving basic notions of space, time, causality, localizability of matter, etc. So it is possible to understand the one-body quantum system without the need for any really striking changes in the overall conceptual structure of physics.

It is only when we try in this way to understand the many-body system that quantum mechanics begins to show the need for a radically novel general conceptual structure. To show how this comes about, let us first consider a two-body system (with two particles of equal mass, m). The wave function, Ψ, depends on the six variables, x_1 and x_2, constituting the co-ordinates of the two particles, as well as on the time, t. Evidently, $\Psi(x_1, x_2, t)$ can no longer be thought of as a field in ordinary three-dimensional space. Rather, it is a function defined in the configuration space of the two particles. And thus, Ψ, has no such a simple and direct physical interpretation as is possible in the one-particle case. Nevertheless, we can still develop an indirect interpretation for it by writing $\Psi = Re^{iS/\hbar}$ and by substituting in the Schrödinger equation for the two-particle system. The result is

$$(6) \qquad \frac{\partial P}{\partial t} + \nabla_1 \cdot \left(P\frac{\nabla_1 S}{m} \right) + \nabla_2 \cdot \left(P\frac{\nabla_2 S}{m} \right) = 0,$$

$$(7) \qquad P = \Psi^*\Psi,$$

$$(8) \qquad \frac{\partial S}{\partial t} + \frac{(\nabla_1 S)^2}{m} + \frac{(\nabla_2 S)^2}{m} + V(x_1, x_2) + Q = 0,$$

$$(9) \qquad Q = -\frac{\hbar^2}{2m}\left(\frac{\nabla_1^2 R}{R} + \frac{\nabla_2^2 R}{R} \right).$$

Evidently, Equation (6) describes the conservation of probability $P = \Psi^*\Psi$ in the configuration space, x_1, x_2 of the two particles. Equation (8) is a Hamilton-Jacobi equation for the system of two particles, acted on not only by the classical potential V but also by the quantum potential $Q(x, x_2, t)$. Thus, if we take into account that in the Hamilton-Jacobi equation, $v_1 = \nabla_1 S/m$ and $v_2 = \nabla_2 S/m$ we have an implicit determination of the meaning of $\Psi = Re^{iS/\hbar}$ (in the sense that R determines both the probability density and the quantum potential, while S determines the mean momenta of the particles).

In an N-body system, we would have $\Psi(x_1, \ldots, x_N, t)$ with $Q = Q(x_1, \ldots, x_N)$. This gives rise to what is called a 'many-body force', i.e. an interaction that does not reduce to a sum of terms, one for each pair (e.g. $F(x_m - x_n)$), but rather one in which the interaction between each pair depends on all the other particles. Of course, such forces do not necessarily go beyond the general conceptual framework of classical physics (e.g. the van der Waals force may in principle be a many-body force, in which the interaction between a given pair of molecules is capable of being influenced by all the other molecules in the system).

However, what is new here are the following two points:

(a) The quantum potential, $Q(x_1, \ldots, x_N)$ does not in general produce a vanishing interaction between two particles, i and j, as $|x_i - x_j| \to \infty$. In other words, distant systems may still be in a strong and direct interconnection. This is, of course, contrary to the general requirement, implicit in classical physics, that when two particles are sufficiently far apart, they will behave independently. Such behaviour is evidently necessary if the notion of analysis of a system into separately and independently existent constituent parts which can conceptually be put together again to explain the whole is to have any real meaning.

(b) What is even more strikingly novel is that the quantum potential cannot be expressed as a universally determined function of all the co-ordinates $x_1, \ldots, x_N$. Rather, it depends on $\Psi(x_1, \ldots, x_N)$ and therefore on the 'quantum state' of the *system as a whole*. In other words, even apart from the point made in (a), we now find that the relationships between any two particles depend on something going beyond what can be described in terms of these particles alone. Indeed, more generally, this relationship may depend on the quantum states of even larger systems, within which the system in question is contained, ultimately going on to the universe as a whole.

If this dependence of the relationships of parts on the state of the whole is what generally prevails, how then are we to understand the fact that in broad domains of physical experience, the world can successfully be treated by analysis into separately existent parts related in ways that do not thus depend on the whole?

To answer this question we first consider the special case in which the wave function is factorizable as a product. In a two-body system, for example, suppose we can write

(10) $\Psi = \phi(\mathbf{x}_1)\phi(\mathbf{x}_2),$

(11) $P = |\Psi|^2 = |\phi_1|^2|\phi_2|^2,$

(12) $Q = -\dfrac{\hbar^2}{2m}\left(\dfrac{\nabla_1^2 R_1(\mathbf{x}_1)}{R_1(\mathbf{x}_1)} + \dfrac{\nabla_2^2 R_2(\mathbf{x}_2)}{R_2(\mathbf{x}_2)}\right).$

In this case, the quantum potential reduces to a sum of terms, each dependent only on the co-ordinates of a single particle, and therefore each particle behaves independently. So as far as *the functioning of the particles is concerned* we can get back the quasi-independence of behaviour of the particles once again. But now, this comes out as a special case of non-independence (i.e. quantum interconnectedness).

In Ref. [8], it was shown in some detail that this theory actually does give a complete and consistent way of understanding the (non-relativistic) many-body system. To help bring out in more detail how this works, let us now look briefly at the EPR experiment. In the original proposals of EPR, one considers a system of two 'non-interacting' particles in a quantum state given by the wave function

(13) $\Psi = \exp\left[ip\dfrac{(\mathbf{x}_1 + \mathbf{x}_2)}{2}\right]\delta(\mathbf{x}_1 - \mathbf{x}_2 - a).$

This is a state in which the total momentum of the system is p, and in which the two particles are separated by the distance, a, which as EPR noted, may be of macroscopic dimensions. If one now measures the position of the first particle, $\mathbf{x}_1$, then it follows that $\mathbf{x}_2 = a + \mathbf{x}_1$. This could perhaps be explained by supposing that $\mathbf{x}_1$ and $\mathbf{x}_2$ were initially correlated in this way before the measurement took place. But then, one could instead have measured p_1 and obtained the correlated result, $p_2 = p - p_1$. However, according to the uncertainty principle, $\mathbf{x}_1$ and p_1 cannot both be defined together. It would follow that somehow the measurement of the momentum of the first particle actually 'put' this particle into a definite state of momentum, p_1, while it 'put' the second particle into a correspondingly definite correlated state of momentum, $p - p_1$. The paradoxical feature of this experiment then is that particle 2 somehow seems to 'know' into which state it should go, without any interaction that could transmit this information.

Let us now consider how all this is understood in terms of the causal interpretation. Firstly, to avoid infinite terms, we will replace the wave function (13) by

$$(14) \qquad \Psi = \exp\left[ip \frac{(x_1 + x_2)}{2} \right] f(x_1 - x_2),$$

where f is a real function, sharply peaked at $x_1 - x_2 = a$. By allowing the peak to grow sharper and sharper, we can approach as near as it is necessary to the wave function (13).

Let us now write down the quantum potential

$$(15) \qquad Q = - \frac{\hbar^2}{2m} \frac{\nabla^2 f(x_1 - x_2)}{f(x_1 - x_2)}.$$

This implies that even when the classical potential vanishes (so that in the usual interpretation of the theory, it is said that the two particles do not interact) there is still a 'quantum interaction' between them, which does not approach zero as $|x_1 - x_2|$ approaches macroscopic dimensions, and which depends on the quantum state of the whole system.

When any property (e.g. the momentum of particle No. 1 is measured, then particle No. 2 will react in a corresponding way, as a result of the interaction brought about by the quantum potential. And so (as shown in some detail in Ref. [8]) it is possible intuitively and imaginatively to explain the correlations of measured properties of the two particles, which have no such explanation in terms of the usual interpretation of the theory.

The above explanation of the experiment of EPR helps us to understand how quantum interconnectedness may come about, even over macroscopic orders of distance. But such interconnectedness shows itself in many other ways which have (like the experiment of EPR itself) been confirmed in actual observation. For example, in the superfluid state of helium, it is well known that there are long range correlations of helium atoms [10]. The wave function for such a state can easily be shown to lead to a quantum potential in which there is in general a significant interaction between distant particles. At high temperatures, such correlations become unimportant, so that helium atoms which are distant from each other move more or less independently, with the result that they undergo random collisions, leading to viscous friction. Below the temperature of transition to the superfluid state, however, the long range interactions implied by the quantum potential give rise to a co-ordinated movement of all the particles (resembling a ballet dance rather than a disorderly crowd of people). And so, because there are no random collisions, the fluid is free of viscous friction (a similar explanation can also be given for superconductivity).

In this way, we see how in certain domains of large scale experience, we can intuitively apprehend the essential quality of quantum interconnectedness, in the sense, for example, that we understand the observable property of superfluidity as the result of the movements of subsystems (atoms) whose relationships (i.e. interactions over long distances) depend on the state of the whole system. Thus, we do not have to restrict ourselves to a purely mathematical treatment, from which one can derive formulae implying the vanishing of the viscosity, without however obtaining a real imaginative insight into why this happens.

3. ON A NEW NOTION OF UNBROKEN WHOLENESS

As indicated in Section 2, the quantum theory understood through the causal interpretation implies the need for a radical change from the classical notion of analyzability of the world into independently existent parts, each of which can be studied in relative isolation, without our having to consider the whole, and which can in turn be put together conceptually to explain this whole. Rather, the basic qualities and relationships of all the 'elements' appearing in the theory are now seen to be generally dependent on the state of the whole, even when these are separated by macroscopic orders of distance. However, when the wave function can be expressed approximately as a product of functions of co-ordinates of different 'elements', then these latter will behave relatively independently. But such a *relative independence of function* is only a special case of general and inseparable dependence. So we have reversed the usual classical notion that the independent 'elementary parts' of the world are the fundamental reality, and that the various systems are merely particular contingent forms and arrangements of these parts. Rather, we say that inseparable quantum interconnectedness of the whole universe is the fundamental reality, and that relatively independently behaving parts are merely particular and contingent forms within this whole.

We can express this view more succinctly by considering the possible meaning of the word 'system'. Now each system may be regarded as constituted of many subsystems while in turn such a system may be considered as a constituent of various supersystems. Thus, we may say that a standard or canonical form of treatment in physics has been to make a distinction that establishes three levels of description

> Supersystem
> System
> Subsystem

Classically, it has been supposed that we can eventually arrive at subsystems (e.g. elementary particles) whose basic qualities and relationships are independent of the states of the systems and supersystems in which they participate. But quantum mechanically, as we have seen, we cannot arrive at such an analysis. Nevertheless, this distinction into supersystem, system and subsystem is still both meaningful and useful. However, its role is not to give an analysis into constituent parts, but rather to serve as a basis of *description*, which does not imply the independent existence of the 'elements' that are distinguished in this description (e.g. as we may describe a ruler as divided into yards, feet, inches, without implying that the ruler is ultimately constituted of separately existent 'elementary inches' that have been put together in some kind of interaction). The distinction into supersystem, system and subsystem is thus a convenient abstraction, which in each case has to be adapted to the actual content of the physical fact.

We emphasize that in this point of view there can be no ultimate set of subsystems nor an ultimate supersystem that would constitute the whole universe. Rather, each subsystem is only a relatively fixed basis of description. Thus, atoms, originally thought to be the absolute and final constituents of the whole of reality, were later found to be only relatively stable units, being constituted of electrons, protons, neutrons, etc. And now, physics seems to be pointing to the likelihood that such 'elementary particles' are also only relatively stable units, possibly being constituted of finer elements, such as 'partons'. However, we are not basing our view here solely on contingent facts of this kind. Rather, what we are proposing is that we be ready to explore a new notion of physical reality, in which we start from *unbroken wholeness* of the totality of the universe. Any attempt to assert the independent existence of a part would deny this unbroken wholeness and would thus fail to be consistent with the proposed notion. To state that a given part is ultimate, in the sense of not being describable in terms of subsystems would likewise still give this part a certain absolute character that would be independent of the whole.

This does not necessarily mean that the subsystems are always spatially smaller than the system as a whole. Rather, what characterizes a subsystem is only its relative stability and the possibility of its independence of behaviour in the limited context under discussion. For example, a crystal can be described as a system of interacting atoms. But it can also be described as a system of interacting normal modes (sound waves). In this latter description, the subsystems are the normal modes. These are spatially co-extensive with the system as a whole. But *functionally*, the normal modes have a relative stability of movement and possibility of independent behaviour which allows them to be consistently regarded as subsystems of the crystal as a whole.

When we turn in the opposite direction toward the supersystems in which a given system participates, similar considerations will hold. Thus, in general, it will not be possible to regard the relationships of systems as being independent of such supersystems. A very simple example of this in physics is provided by the process of measurement in quantum mechanics. If we regard a 'particle' as 'the observed system', then we cannot properly understand their relationships, except in the context of the overall experimental situation, created by the 'observing apparatus' which, together with the observed object, has to be regarded as a kind of supersystem (in Ref. [8], the measurement process is treated from this point of view in some detail).

There is a considerable similarity here to Bohr's point of view. Thus, Bohr emphasizes the wholeness of the form of the experimental conditions and the content of the experimental results[5]. But Bohr also implies that this wholeness is not describable in intuitive and imaginative terms because (in his view) we must use *classical language* and *classical concepts* to describe our actual (large scale) experience in physics, and these evidently contradict the wholeness that both Bohr and we agree to be necessary.

In our approach, we do however, give an intuitive and imaginative description of such wholeness. We do this by dropping the notion that actual large scale experience in physics has to be described solely in terms of classical concepts. Rather, even when we describe such experimental results, we have explicitly to incorporate these into the language of supersystem, system and subsystem. In this way, we will imply that whatever is observed at the large scale level is only a relatively stable system, in which the relationships of the subsystems may depend significantly on the state of the system as a whole.[1] Thus we differ from Bohr in giving an intuitive and imaginative description of our large scale experience, which does not contradict the wholeness implied by the quantum properties of matter.

In this way we are able to have not only unbroken wholeness in the *context* of physics, but also we can have such unbroken wholeness in the *form* of description, in the sense that the one form of supersystem, system and subsystem is valid for the whole field of physics, large scale and small scale. Subsystems will then generally depend intimately on the systems in which they participate, which will in turn depend on supersystems, etc., ultimately merging with the unknown totality of the whole universe, with no sharply delineated cuts or boundaries. In principle, this includes even the observer. But in typical cases, the observer himself can be treated as functionally independent of what is observed (i.e. the observed object is taken as the system, its deeper constitution as the set of subsystems, and the overall environment, including the observing apparatus, as the supersystem). So without implying the breaking of wholeness, we are able to make a certain relatively independent physical context stand out 'in relief' against a background (including the observer) that is not important for its function or behaviour. The whole is thus always implicitly present in every description. In different contexts, the fundamental independence of what we are studying may cease, and we may then have to go back to the whole, to come to a new act of abstraction, making a different structure of supersystem, system and subsystem stand out 'in relief'.

It is clear then that *wholeness of form* in our description is not compatible with *completeness of content*. This is not only because subsystems may eventually have to be regarded as constituted of sub-sub systems, etc. It is also because even supersystems will ultimately have to be seen as inseparable from super-super systems, etc. This form of description cannot be closed on the large scale, any more than on the small scale. Thus, if we supposed that there was an ultimate and well defined supersystem (e.g. the entire universe) then this would leave out the observer and it would break the wholeness, by implying that the observer and the universe were two systems, separately and independently existent. So, as pointed out earlier, we do not close the description on either side, but rather, we regard the supersystems and subsystems as ultimately merging into the unknown totality of the universe.

It follows then that because there is no ultimate description, each level makes an irreducible contribution to the content of the description. Thus, when we describe a superfluid as constituted of helium atoms, we have not completely reduced it to a set of helium atoms because the interactions of these atoms are still determined by the state of the whole system. Similarly, the helium atoms are made up of electrons, protons and neutrons, but the behaviour of these depends on the state of the helium atom as a whole. If the elementary particles have a finer constituent (e.g. partons) the behaviour of these latter will still depend on the state of the particle as a whole. So each level in any description makes an irreducible contribution to the overall description and to the expression of the overall law.

It is clear then that incompleteness of content of a theory is necessary for wholeness in form. A theory that is whole in form may be compared with a seed that can grow in an indefinite number of ways, according to the context in which it finds itself. But however it grows, it always produces a plant in harmony with the environment, so that together they constitute a whole. Clearly, this is possible only because the articulation of the plant is not pre-determined in detail in the seed. On the other hand, any theory that pretends to completeness of content must close itself off from the unknown totality in which all ultimately merges so that it will eventually give rise to fragmentation in form.

Of course, we have thus far been discussing wholeness of form in terms of the causal interpretation of the quantum theory. As indicated in Section 1, we are quite aware of the limitations and unsatisfactory features of the interpretation. Nevertheless, we feel that it has served a useful purpose, by leading us to an intuitive and imaginative understanding of the unbroken wholeness of the totality of existence, along with the possibility of abstracting specified fields of study having relative functional independence, as described through the distinctions of system, supersystem and subsystem. Our attitude is that we can sooner or later drop the notion of the quantum potential (as we can drop the scaffolding when a building is ready) and go on to radically new concepts, which incorporate the wholeness of form – which we feel to be the essential significance of quantum descriptions. This implies that we have to go deeply into all our basic notions of space, time and the nature of matter, which are at present inseparably intertwined with the idea of localizability, i.e. that the basic form of existence is that of entities that are located in well-defined regions of space (and time). We have instead to start from non-locality as the basic concept, and to obtain locality as a special and limiting case, holding when there is relative functional independence of the various 'elements' appearing in our descriptions. This means that our notions of space and time will have to change in a fundamental way.

At present we are working on these lines. Some progress has been made [11, 12] and, in the next section of this paper, some of the further possible directions of work in which we are now engaged will be discussed briefly.

Before going on to this, however, we point out that the form of wholeness (i.e. supersystem, system, subsystem, with parts dependent on the state of the whole) is relevant in fields going far beyond physics. Thus, for example, individual human beings may be considered as subsystems in a system consisting of a social group, which is in turn part of a supersystem, consisting of a larger social group. Evidently, the relationships of any two individual human beings depend crucially on the state of the immediate social group to which they belong, and ultimately on that of the larger social group.

Similarly, the interactions of any two cells in the body depend on the state of the whole organ of which they are a part, and ultimately on that of the organism as a whole. Likewise, such a description holds for mental phenomena (e.g. the relationships of the two concepts depend crucially on broader concepts, within which they are comprehended, etc.). In this way we see that there is accessible to us a very wide range of direct intuitive experience in the form of wholeness. What the quantum theory, as understood through the causal interpretation, shows is that this form is appropriate not only biologically, socially and psychologically, but also, even for understanding the laws of physics. And so we are able to comprehend the whole world in all of its aspects through the one universal order of thought, thus removing an important source of fragmentation between physics and other aspects of life.

4. ON A POSSIBLE EXTENSION OF THE NOTION OF UNBROKEN WHOLENESS TO RELATIVISTIC CONTEXTS

Thus far, we have restricted ourselves to a non-relativistic context, so that the instantaneous interactions of distant particles implied by the quantum potential do not give rise to any fundamental difficulties of principle. There does not exist as yet, however, a consistent relativistic quantum theory, with a generally satisfactory physical interpretation. At present, relativistic quantum mechanics is a set of fragmentary algorithms, each working in its own domain, but there is no clear theory of how these algorithms form a whole body of theory that would be related in some definite way to experiment (e.g. there is no consistent relativistic theory which includes a theory of measurement). So all we can do now is to see if we can obtain some kind of hints or indications of what is needed in a relativistic context, by looking at this problem through the intuitive notions given here.

First of all, we point out that in the theory of relativity, the concept of a signal plays a basic role in determining what is meant by separability of different regions of space.[2] In general, if two such regions, A and B, are separate, it is supposed that they can be connected by signals. Vice-versa, if there is no clear separation of A and B, a signal connecting them could have little or no meaning. So the possibility of signal implies separation, and separation implies the possibility of connection by a signal.

Can the quantum potential carry a signal? If it can, we will be led to a violation of the principles of Einstein's theory of relativity, because the instantaneous interaction implied by the quantum potential will lead to the possibility of a signal that is faster than light. However, the mere fact of interaction does not necessarily give rise to the possibility of carrying a signal. Indeed, a signal has, in general, to be a complex structure, consisting of many events, that are ordered in definite ways. Or, in terms of our language, each signal is a super-system of events, while each event is in turn a system of sub-events.

It is not at all clear that the action of the quantum potential on a particle would have the regular order and complex stable structure of super-events, events and sub-events needed to carry a meaningful signal. And if such a signal could not be carried by the quantum potential, then the basic principles of relativity theory would not necessarily be violated. So one avenue of enquiry is to try to find out to what extent the quantum potential can carry a signal. At present, the answer is, of course, not known.

For example, as pointed out in Section 3, in the experiment of EPR, an observer, A, can measure either the position or the momentum of particle 1, and the corresponding property of particle 2 will be immediately determined in a properly correlated way. From a measurement on particle 1, the observer, A, knows the corresponding property of particle 2. But an observer, B, near particle 2, cannot learn anything from this that he did not know before. For example, if he measures the momentum of particle 2, he will get a definite result, but he will have no way of knowing that the observer, A, has helped to produce this result, (unless information is exchanged between observers in some other way). This follows, because the events in a measurement of momentum do not have enough structure to carry to observer B any information as to what has been measured by observer A, and also because any event observed by B could have occurred equally well with or without the action of observer A. So it is quite possible that the instantaneous interactions implied by the quantum potential do not violate the basic principles of relativity theory.

A closely related point is that the wave function of a many-body system is written as $\Psi(x_1, \ldots, x_N, t)$ with a single time, t, for all the particles. In other words, all the particles have to be considered together at the same time, if they are to be treated as a single system.

If one does not do this, one does not in general obtain the correct experimental results (e.g. for stationary states). And evidently, the instantaneous interaction of the quantum potential, which depends on $|\Psi(x_1, \ldots, x_N, t)|$ is a consequence of the fact that all particles are considered at the same time, in the definition of the quantum state.

Clearly, there is something deeply non-covariant in this procedure. We could, however, remove this non-covariance by introducing here the new concept of a particular time frame associated to each stable many-body system in a stationary (or quasi-stationary) state. Systems moving at different speeds would have different time frames related by a Lorentz transformation.

In this way, we have extended our descriptive form of super-system, system and subsystem by proposing that the relevant time order for subsystems (e.g. atoms) in a system (e.g. a solid block of matter) depends on the overall state of movement of the whole system, and is not (as implied in classical and commonsense notions of time) fixed in a way that is independent of the state of the whole system. Moreover, general relativity shows that this time order depends even on the states of supersystems (e.g. on the gravitational fields of the surrounding matter and of the cosmos as a whole).

The time order that belongs to a system is not only relevant for *descriptive purposes*. It is also relevant dynamically in the sense that each system has a certain characteristic time frame, within which 'instantaneous' interactions between subsystems are possible. But, as indicated earlier, this theory may still be covariant, firstly because the same laws hold for all systems, regardless of their speeds, and secondly, because these 'instantaneous' interactions may not be capable of carrying a signal.

In this way, the concept of time (and ultimately of space) may be enriched, not only to fit the principles of relativity, but also to harmonize with the general spirit underlying a relativistic approach. For the relevant time order of a subsystem is now relative to the system and supersystem within which the subsystem participates. This makes the time order well defined; i.e. non-arbitrary, without however implying a universally fixed and absolute order of time.

We see then that with the aid of an intuitive understanding of the quantum mechanics, we are led to the possibility of new ways of looking at both quantum theory and relativity theory, along with new ways of bringing them together in terms of novel concepts of time (and space) order. Work is now going on along these lines, on which we hope to report later.

Birkbeck College, University of London

NOTES

* This paper has already been published in *Foundations of Physics* **5** (1975), 93–109.
¹ E.g. as we have suggested in Section 2 for the case of superfluid helium.
² For a more detailed discussion of this point, see Ref. [11].

BIBLIOGRAPHY

[1] Einstein, A., Podolsky, B., and Rosen, N., *Phys. Rev.* **47** (1935), 777.
[2] Schrödinger, E., *Proc. Camb. Phil. Soc.* **31** (1935), 555.
[3] Bell, J. S., *Physics* **1** (1964), 195.
[4] Freedman, S. J. and Clauser, J. F., *Phys. Rev. Letters* **28** (1972), 938.
[5] Bohr, N., *Atomic Theory and the Description of Nature* (Cambridge University Press, 1934) and *Atomic Physics and Human Knowledge* (Wiley, New York, 1958).
[6] Jauch, J. M. and Piron, C., *Helv. Phys. Acta.* **42** (1969), 842.
[7] Finkelstein, D., *Phys. Rev.* **184** (1969), 1261 has developed a view of quantum logic which is relatively imaginative in its point of departure. However, ultimately, his basic concepts are mathematical rather than intuitive, so that he is, for example, unable to bring out the sort of points that we emphasize in this article.
[8] Bohm, D., *Phys. Rev.* **85** (1952), 166, 180.
[9] Bohm, D. and Vigier, J.-P., *Phys. Rev.* **96** (1954), 208.
[10] Vinen, F. W., *Reports on Progress in Physics* **31** (1968), 61.
[11] Bohm, D., *Foundations of Physics* **1** (1971), 359; and *Foundations of Physics* **3** (1973), 139.
[12] Bohm, D., Hiley, B. J., and Stuart, A. E. G., *Int. J. of Theoretical Phys.* **3** (1970), 171.

R. LESTIENNE

FOUR IDEAS OF DAVID BOHM
ON THE RELATIONSHIP BETWEEN
QUANTUM MECHANICS AND RELATIVITY*

ABSTRACT. This paper deals with some remarks of D. Bohm on the interpretations of quantum mechanics in the view of opening a debate on the relationship between quantum mechanics and relativity. Such problems as those of signal, speed of light, reference frames, and the fourth uncertainty relation are evoked.

We do not intend here either to explain the whole of D. Bohm's interpretation of quantum mechanics, or even to discuss in detail only a few of its aspects. Because on the one hand we do not consider ourselves competent to do so and on the other, Bohm explains his own ideas in this volume. Our ambition is more modest. It is to open a debate on the relationship between quantum mechanics and relativity, making use of some remarks Bohm made about quantum mechanics, at least as I understood them.

At the outset, it must be said that the question here will be neither to criticize the relativistic quantum theories that may exist yet, nor to deny the successes of relativistic quantum mechanics or those of quantum electrodynamics, but to recognize the limits of these theories. We must recognize the difference between *adapting an existing quantum theory* with the requirements of Lorentz invariance, and *elaborating a theory which is both relativistic and quantal*, which would demand an analysis of its concepts profound enough to bring harmoniously together the basic concepts of quantum mechanics and those of relativity. Of course, up to now, only the first of these two requirements has been satisfied; but the successes obtained and the difficulties encountered can act as a guide in approaching the second requirement.

On the contrary to knit together the basic concepts of quantum mechanics and those of relativity in the same theory appears to be an extremely arduous programme that has not yet been achieved. The basic concepts of relativity are those of signal, event and invariance of the observations realized in reference frames that are in relative motion one to another; while the concepts of quantum mechanics, at

J. Leite Lopes and M. Paty (eds.), Quantum Mechanics, a Half Century Later, 227–235. All Rights Reserved

least as understood by D. Bohm and those who adopt Bohr's point of view, are essentially those of *non-separability* and of *quantum jumps*, that is to say the existence of essential discontinuities in the evolution of quantum systems. As Bohr often used to insist, there is a basic non-separability between the quantum object to be studied and the instrument for its observation. One cannot speak of quantum system, at least in classical terms, if no measurement apparatus is associated with it. Even more basically, the replacement, in quantum mechanics, of the punctual point, which represents in classical mechanics the dynamical state of a particle by a phase extension cell whose volume is $\hbar^3$ and whose form moreover depends explicitly on the experimental set-up used, shows the non-separability between the quantum object and the instrument (in Bohr's meaning) in relation to his complementary principle. Non-separability is much more general than that. It is linked to the numerous debates to which the Einstein, Podolski and Rosen paradox has given rise. Professor d'Espagnat showed us that this non-separability seems to be inscribed in a demonstration that would imply only very general axioms, even more general than the specific axioms of such and such form of quantum mechanics.

The other fundamental question to consider is that of quantum discontinuities. One has, for example, the fact that no atoms exist in the process of decaying, only atoms before and after their decay, or the fact that in atoms there do not exist any electron being, even momentarily, in a state of energy intermediate between the first excited level and the ground level, but there are only electrons in the ground level or in the excited states. Therefore in quantum mechanics there is a fundamental discontinuity that must be taken into consideration in order to elaborate a physical theory that is equally quantal and relativistic.

These two considerations of non-separability and existence of quantum jumps lead Bohm to an analysis in which he distinguishes between what he calls the *explicit orders*, and the *implicit orders*. Nature, he says, admits very general structural relations, that may not necessarily and immediately be placed into the spatio-temporal framework. The aim of physical theory is to describe such structural relations; but our mode of existence, our condition of observers and our mental structure do not allow us to grasp directly these abstract relations. We can understand only certain types of order, the explicit orders. But these are revealed to our senses or to our instruments only through certain of well determined experimental configurations, such as the taking of a photograph at some given place of some given experimental set-up. These explicit orders are, of course, linked to the fundamental structural relations, but they only present a certain aspect of these relations which are themselves actually so rich that many orders, alongside the explicit orders, remain implicit.

Thus each experiment that reveals an explicit order, represents a way of seeing nature, but many other ways could be imagined; in other experiments, other orders would become explicit, the preceding order remaining latent (implicit). According to D. Bohm, everything that for us stems from to the space-time organization of the world (an organization which is at the basis of relativity), refers to a special type of explicit orders. They bring into play standards of length and time, instruments for the comparison of observable, various mental structures, and various psychological attitudes founded upon an analysis of the world in terms of separate objects. Like all the explicit orders, they are no more than a partial reflection of the fundamental structural relations that exist in nature with which physical theory must try to get to grips.

To make things somewhat clearer, let us now consider a hologram. It presents itself to the naked eye as a structure stretching in space, involving a complex of apparently disordered dark spots. But actually there exists an order in this network, for if the hologram is exposed to the light of a laser of convenient frequency, at once one can see the structural relations that link the spots in the hologram. Thus one now perceives as an explicit order what was before only implicit and non-perceived when the hologram was observed by the naked eye. The laser revealed this order, made it pass from the state of an implicit order to that of an explicit one. In this analogy one can see on the one hand the idea of a non-separable universe in which everything is in the whole and vice-versa, the explicit order being better perceived in proportion to the fraction of the hologram simultaneously observed. Also there is the idea that a given explicit order describable by certain concepts can be made to appear only in well-determined experimental situations. In other experiments, for instance, either the concept of wave or the concept of particle will be appropriate for the description of the explicit order.

Let us now consider the trajectory of a microscopic particle that

might be observed in the instruments of high energy physics labora-
tories. In reality a trajectory observed in these instruments is never
continuous: it is made of a succession of spots, for instance bubbles
in a bubble chamber or ionized atoms in an emulsion. When these
spots are looked at from far enough, i.e. globally, we can perceive at
once a structure, an explicit order, which is called the trajectory of
these particles, but which is actually a big abstraction from the
physical reality that consists of more or less disordered spots. In this
case the fundamental reality is not the trajectory, but the existence of
these discontinuous spots, or, to go further, the existence of the
particle or the state vector that describes it. The points of view are to
a certain extent relative, since when the explicit order is considered
as fundamental, one speaks of a trajectory, whereas when the quantum object is considered fundamental the 'trajectory' order will
become implicit in relation to the former.

    The ideas of Bohm on the fundamental relations between quantum
physics and relativity with which I chose to open the debate are the
following ones: *firstly*, the notion of signal is seriously limited. It is
indeed an explicit order, but at the same time it is a means of
conveying an instruction from one place to another, step by step,
through the continuous propagation of an energy (we use here the
word 'order' in connection with the word 'structure', but the word
'instruction' in connection with relativity, instead of 'order', so as to
avoid any confusion). From the moment when non-separability holds,
in certain situations the distinction between the transmitter of the
instruction and the receiver will surely be quite difficult. It seems
Bohm considers the signal as a part of the world of explicit orders, as
described above, related to the macrocosm and to our point of view
on the world, but without any intrinsic significance at a microphysical
level. For instance, the propagation of instruction cannot be seen
naturally in the experiment described previously in which a particle of
spin 0 decays into two particles of spin $\frac{1}{2}$, i.e. the propagation of an
instruction from one particle to another to tell it what to do in such
and such an experimental situation. This non-propagation of an
instruction, in the relativistic sense, is now being tested experimentally.

Bohm's *second* idea treats the question of the speed of light as an upper bound on movement. This limit concerns only transmitting instructions; it does not concern the conveyance of objects or structures if these cannot be used to transmit a signal. Movement as it appears in the orthodox description of quantum mechanics (Bohr's point of view) surely cannot be limited to the speed of light. Let us consider the motion of a particle as described above. The spots materialize from successive measurements of the particle position at very near instants. These measurements do not show any continuity between themselves, thus depicting the existence of quantum discontinuities, which we said are the basic data of microphysics. The limit of $\Delta x / \Delta t$ of the differences of the positions at two successive instants cannot be finite when Δt tends to 0, because then the result would be to make a simultaneous measurement of position and of speed, which from the principles of quantum mechanics is forbidden.

In the example of Brownian motion, which might be considered as the archetype of motion in quantum mechanics, the successive position of the colloidal particle may therefore move at instantaneous speed, greater than the speed of light. This is not in contradiction with the basic axiom of relativity because we deal here with a movement that cannot convey a signal. It is wholly unpredictable, in fact, that the brownian particle, which had appeared at a certain position at the instant t should appear in another position at the instant t': this

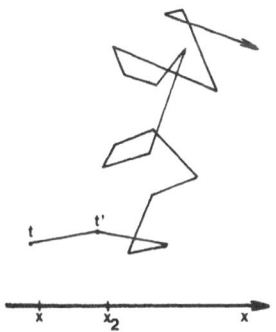

results from the reduction of the wave-packet. This Brownian movement is therefore wholly unordered, and does not allow us to tell an observer, located in X_2 for instance, to set in motion a certain apparatus when the particle reaches it, because it cannot be predicted with certainty whether the particle will pass in X_2 at the moment t', or at t_2' or at any other later instant. But on the other hand, we can compute the average time for the particle to propagate from X to X_2. This average time of course gives an average speed that does not exceed the speed of light. We can in practice rely on this average speed to propagate an instruction.

The *third point* concerns the necessary existence (according to Bohm) of a material reference frame to describe quantum systems. In classical mechanics, if not in its internal logic, at least in its historical development, the notions of absolute space and absolute time formed a necessary framework for the description of natural phenomena. Relativity tried to get round this constraint; but is it actually possible? Electromagnetism provides no explanation of the stability of matter. (Here we make an allusion to the fundamental discussion around this point in the 20's, which motivated the invention of quantum mechanics.) Consequently the rods and the clocks needed to introduce the concepts of space and time in relativity cannot be built if we confine ourselves strictly to classical physics and classical electromagnetism. Indeed the notion of stability does not really appear anywhere except in quantum mechanics. Hence quantum mechanics is necessary to account for the stability of matter and for the introduction of the standards of length and time that make it possible to know what we are talking about when we speak of the position and duration of a certain phenomenon.

What, then, is a material reference frame? It is in practice a huge collection of atoms in their stationary states, which makes possible, since the atoms are in stationary states, an adequately precise definition of a standard of time and a standard of space. But here there arises the difficulty similar to the one encountered in classical mechanics. A system that is stationary in a certain frame is not so for another frame moving with respect to it. This can easily be understood; in classical mechanics the orbit of a planet is stationary only in relation to a fixed frame relative to the focus of its motion; in quantum mechanics, a stationary system composed of waves propagating at equal and opposed speeds, does not constitute a system of waves propagating at equal and opposed speed in a frame that is in motion in relation to the first.

Therefore a privileged material frame is needed to shape the necessary referential system in relation to which the state of a given atom may be located and described by means of quantum mechanics. This material system is formed by the experimental surroundings, the measurement apparatus and ourselves, in short a huge collection of atoms in stationary or quasi-stationary states. It is the very existence of this huge collection of atoms in stationary states, that permits one to introduce the explicit structural relations of the space-time type, allowing him to describe the result of the experiments. Since the condition of stationarity of the atoms is not maintained in every other reference system moving in relation to this privileged reference frame, the former explicit orders become, with respect to moving reference systems, implicit orders, i.e. hidden orders that cannot be described by means of the same concepts (even though they do not lose, for all that, their effective reality).

Lastly, the *fourth idea* that I intended to mention briefly, concerns the fourth uncertainty relation. Many thought experiments shows that this fourth relation is closely related to the first three ones of the $\Delta x \, \Delta p \geqslant \hbar$ type. For clarity, recall the thought experiment that Einstein imagined in order to refute the fourth relation, and to which Bohr tried to answer at the Solvay Conference in 1930. A box, with a clock that controls a diaphragm, is hung to a *dynamometer*. At the very moment when the clock rings – a moment that has been determined beforehand – the diaphragm lets out a certain quantity of light energy. By weighing very precisely the box before and after the operation, the quantity of energy that went out at that instant when the diaphragm opened can be found. From this, Einstein argued that the fourth relation must be false. Bohr's very simple refutation takes its argument from the existence of an uncertainty relation of the index position of the box $\Delta z \, \Delta p_z \geqslant \hbar$. By means of a simple argument of general relativity

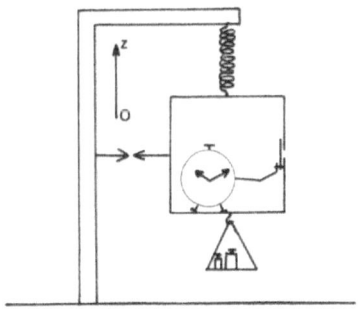

about the time recorded by the clock, it can be shown that $\Delta z \, \Delta p_z \geqslant \hbar$ implies $\Delta E \, \Delta t \geqslant \hbar$. If in Bohr's mind, as well as in Bohm's, these relations of uncertainty must not be interpreted as giving the limits of the precision with which the position and the momentum of the index can be measured, but as giving the limits of relevance of the *words* position and momentum when the quantum system constituted by the box and the index is described, then it follows obviously that we must also speak of an imprecision in the meanings of the concepts of energy and time if we refer to this quantum system.

This is but another argument to illustrate the subtle relation existing between the classical and relativistic framework for the description of phenomena – which uses a continuous space-time and gives rise to a description of an explicit, macroscopic and anthropomorphic order – and the basic order of quantum mechanics which is implicit with respect to the former. The fuzziness we speak of expresses the fact that there does not exist any simple one-to-one relation between the causality parameter t of quantum mechanics and the classical t parameter; the former describes the 'successive' states of the system composed by the box and the pointer, the latter allows the description of the succession of (macroscopic) observations made in a (macroscopic) material reference system external to the box. To pass from one order to the other is to pass from an implicit order to an explicit one, and to operate a *metamorphosis* as Bohm calls it.

[During the short debate that followed, Professor J. M. Jauch pointed out the great similarities existing between the ideas of Bohm and Bohr, especially due to the part played in both approaches by the notions of non-separability and complementarity. Mrs Mugur-Schachter laid stress on the specificity of Bohm's approach, in particular the very profound analysis of the notion of order in physics fundamental to his approach. Some speakers indicated their disagreement about the analysis of movement in terms of Markovian processes, and emphasized the fact that the fourth – so-called uncertainty – relation, does not have the same status as the first three in quantum mechanics.

It must be noted here that beyond any difference of language between David Bohm and Niels Bohr, the breaking point between them lies undoubtedly in their attitude towards the future of quantum mechanics. Bohr proclaimed the completeness of quantum mechanics and the categorical impossibility of speaking of the properties of reality in classical (according to him the only conceivable) language. Thereby he rejected any possibility of progress in the formulation of quantum theory, whilst Bohm believes in this progress. According to the latter, progress will necessarily pass through a new mathematical definition, free from the space-time framework and through the invention of a new non-analytical language, that will be better adapted to the description of quantum reality.]

NOTE

* Paper translated from the French by Yves Paty.

R. DAUDEL

THE ROLE OF QUANTUM MECHANICS IN THE SET-UP OF A MATHEMATICAL GOVERNMENT AMONG MOLECULAR POPULATIONS

ABSTRACT. This paper deals with the problem of the role of quantum mechanics in the analysis of the Universe at a molecular scale. QM has permitted to set up a mathematical government among molecular populations. It is shown, in particular, how QM has replaced the vague chemical concept of bond by a more precise mathematical tool: this question is referred to the loge theory.

1. ABOUT SCIENTIFIC DISCOURSE

Professor Wheeler has pointed out that the Universe has given rise to an observer and the observer gives a sense (meaning) to the Universe. Here, there exists an extensive programme for philosophical meditation. For instance, one could explore this assertion by developing, according to the Heideggerian proposition, an ontological phenomenology starting from an hermeneutic of the 'dasein'[1]. But one can also build numerous scientific discourses about that immense problem. It would like to do this and consider the role of quantum mechanics in the analysis of the Universe at a molecular scale.

Perhaps, it would not be unhelpful to recall the constraints which characterize such types of scientific discourses. The statement, being in some language, must be in accord with the syntax of this language. The logico-mathematical rules which can easily be formalized must encompass its architecture. The discourse should be consistent with the principles of physico-mathematical doctrines and linked to laboratory practice in some way and it is here that the problem of Popperian falsification resides[2].

When such kind of logical discourse is built around some phenomenon,[1] the scientist considers that he possesses a certain amount of understanding of the phenomenon. The power of prediction of science depends on the fact that an analogous discourse may be built on laboratory operations not yet realized, which would describe the results of these operations. From this arises the possibility of confir-

J. Leite Lopes and M. Paty (eds.), Quantum Mechanics, a Half Century Later, 237–244. All Rights Reserved
Copyright © 1977 by D. Reidel Publishing Company, Dordrecht-Holland

mation, the risk of falsifications – one of the motors of the evolution
of scientific knowledge.

If we hope to understand scientifically Wheeler's conclusion beyond the lithospheric field, we should discuss the origin of life, the
evolution of the species and the appearance of memory, consciousness and evidently of speech.

From the molecular view-point, it would be necessary to evoke the
mechanism of formation of the pre-biological systems, the
phenomena of macromolecular duplication, the mutation mechanism,
the elements of a molecular theory of sensations and of memory.
Obviously, it is not possible to analyse, in detail, all these problems in
this brief account. We would like to show only how quantum
mechanics played its role in the setting up of a mathematical
government among molecular populations. This formulation helps one
to explain all molecular phenomena; in other words we can think in
terms of molecules.

2. FROM QUANTUM MECHANICS TO MOLECULAR STRUCTURES

Since every one of us has his own point of view on quantum
mechanics, I would now like to explain mine. With the laboratory, it
is useful to associate the vectorial space R^3. If that laboratory
contains n particles, that is n objects susceptible to produce quasi-
point events, it is normal to introduce the space R^{3n}, or R^{3n+1} if one
includes time. A probability field on R^{3n+1} would correspond to the
occurrence probabilities of an event on R^3 and Professor Jauch first
showed that a natural way to represent such probabilities is to employ
the self-adjoint operators. It is therefore normal that quantum
mechanics should establish, from the functions defined on R^{3n+1}; an
Hilbertian triad $H \subset \mathscr{L}^2 \subset H^*$, where H is a nuclear space, $\mathscr{L}^2$ is the
space of square-summable functions (and also the complement of H
with respect to ordinary scalar product) and H^* the dual of H.

It is well-known that the self-adjoint operators defined therein
possess a complete spectrum.

The connection between this mathematical structure and the
laboratory operations resides, as one knows, in the two principles.

To each 'observable' (i.e. defined by a series of theory-experimental
operations), we associate, following certain rules, a self-adjoint

operator. Each measurement of this magnitude is an eigenvalue of the operator. The occurrence probability of the measurement is related to the projection of the vector representing the particle-system onto the eigenvector associated with this eigenvalue. With the sum of two observables is associated the sum of the corresponding operators. With the product of two observables is associated the product of the operators. With a set of the series of theoretical-experimental operations (observables), provided with two internal composition laws, quantum mechanics associates a set of series of rational operations (the operators), provided also with two such laws. The zone of falsification appears here as a kind of isomorphism.

It seems to me that in this presentation the problem of hidden variables is not so important. However if it does become so, I would like to treat it in the following way. To me, the existence of a trajectory in itself is not a scientific problem but rather a philosophical question (ontological).

Scientifically, it seems to me that only the problem of co-existence is posed. Does a concept contribute or appear to be absolutely foreign to the conceptual structure which would be necessary for me to go from principles to laboratory? Hence, one is led to distinguish three types of possible theories: those where the trajectories may be imagined and calculated from experimental information, those where the trajectories may be imagined but not calculated (hidden parameters), and those which are incompatible with the concept of a trajectory.

The trajectories exist in the first two groups of theories but do not exist in the last group of theories. From the point of view of the application of wave mechanics the concept of trajectory is simply very often useless. For instance, if it is necessary to calculate the energy of a molecule in its ground state, one finds the lowest eigenvalue of the energy calculated in this way for a hydrogen molecule H_2 is 4.7467 eV [3]. The experiment gives: 4.7466 ± 0.0007 eV. This extraordinary agreement is not an isolated fact. Although the solution of the equations are obtained only with less precision in the case of bigger molecules, the energy values are still usually obtained to within a thousandth or several thousandths of the experimental value.

One can, therefore, attach some importance to the eigen-functions corresponding to such values and can thus obtain valuable information, as we shall see next, about the molecular structures.

3. QUASI-CERTAINTY METHOD AND THE MATHEMATISATION OF THE CONCEPT OF CHEMICAL BOND

The notions of chemistry are central to the molecular language. Chemists, physical-chemists and biochemists have accumulated a large amount of data the interpretation of which shows the presence in molecules of objects more or less transferable from one molecule to another. These are called bonds. A chemical reaction is generally considered to be a reorganisation of bonds following a collision:

$$A + B - C \rightarrow A - B + C.$$

Does quantum mechanics permit one to replace that vague concept by a more precise mathematical tool? The loge theory [4], which we shall now state gives a positive response to this question. Moreover, it offers an example of a more general method, applicable to all probabilistic physics, the so-called quasi-certainty method. Let us consider the helium atom in its first triplet state. The old theory of Bohr would consider the two electrons moving around the nucleus one on a K orbit and the other on an L orbit. Quantum mechanics does not allow us to distinguish the K and the L electron.

To simplify, let us replace the nucleus by a fixed charge and let us consider a partition of space in two volumes V_A and V_B. We can define three electron events:

(1) the two electrons 'are' in V_A;
(2) they are both in V_B;
(3) one of them is in V_A and the other in V_B.

The wave function leads to the definition of an occurrence probability for each event. Accordingly, the probability p_3 of the event 3 is written in the form:

$$p_3 = 2 \int_{V_A} dv_a \int_{V_B} dv_b |\Psi(M_a, M_b)|^2.$$

If the boundary between V_A and V_B belongs to a family of surfaces depending on a set $\{\lambda\}$ parameters, each probability p_i is a function of $\{\lambda\}$. The search for the maximum possible information on the localisability of electrons yields a boundary which minimises the missing informations:

$$I = \sum_i p_i \log_2 p_i^{-1}.$$

One finds accordingly, a sphere of radius 1.7 a_0 centered on the fixed charge. Then, the event 3 represents a probability of 93%.

Moreover, the analysis of the function I shows that it cannot be small if one of the events (said to be a dominant event) does not become much more probable than all other events.

The minimisation of I reduces ultimately to the description of a phenomenon studied in such a way that the occurrence of an event becomes a quasi-certainty. Here we find a particular example of an extremely fruitful approach, denoted by the name of quasi-certainty method. It consists in approaching the conditions of classical mechanics (the domain of certainty) by defining particular physical objects described by highly probable values. Hence one can think about the phenomenon by neglecting the other values in the first approximation.

It is said that the sphere of radius 1.7 a_0 corresponds to *the best partition in loges* of the helium state studies. If one desires, one can call loge K the central sphere and loge L the rest of the space. The dominant event is therefore characterised by the presence of an electron in the loge K and an electron in the loge L. It is clear that these characterisations by the different regions of the space and not by the electrons, help us in eliminating the contradiction with the indistinguishability. Figure 1 shows good partitions into loges for various atoms. By dividing the space associated to each loge by the number of electrons during the dominating event, one measures the vital space v occupied by each electron during its presence in the loge. One can also calculate the average electrical potential p, which it experiences in that loge, originating from nucleus and other electrons. It is seen that:

$$p^{3/2}v = \text{constant}$$

for all the atoms and all the loges. *There exists a kind of law of Boyle Mariotte between size of the electron and electrical pressure which compresses them.*

The loge theory may be applied to the molecules. If we consider the BH molecule in the approximation of Born and Oppenheimer (the nucleus replaced by charge fixed at the 'equilibrium' position), we undertake a problem of 6 electrons (6 bodies). Let us select the best

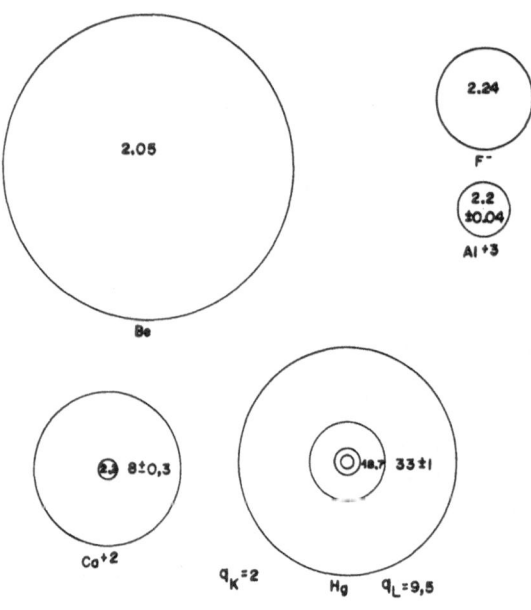

Fig. 1.

way of partitioning in three 'loges', one being a sphere of radius R centered on the boron nucleus, the other two separated by a cone of angle θ with BH direction as the axis (Figure 2). The missing information is minimised for $R = 0.7\ a_0$ and $\theta = 73°$. The dominating event corresponds to two electrons in each '*loge*'. The sphere will be called the core loge of boron atom. The interior of the cone constitutes the BH bond loge and the exterior the lone pair loge.

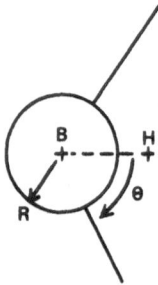

Fig. 2.

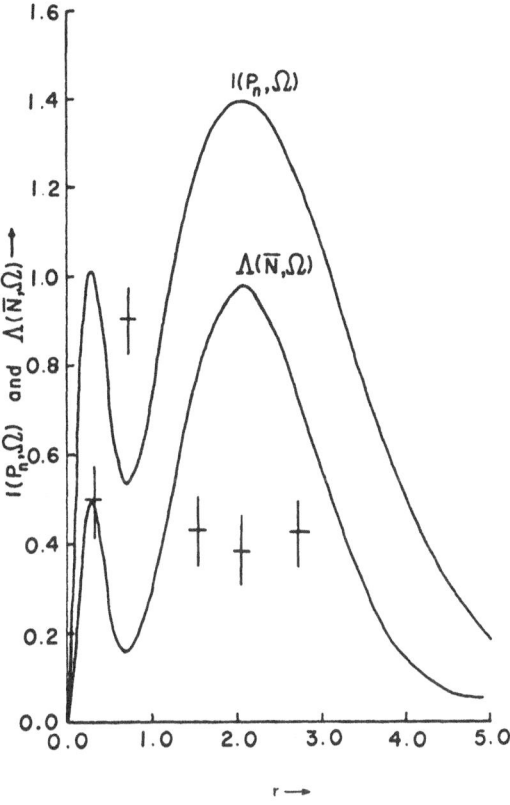

Fig. 3.

Figure 3 shows that the fluctuation Λ of the number of electrons in the loges is minimised at the same time as the missing information. A chemical bond thus becomes a region of the molecular space whose number of electrons fluctuates very little. Moreover, it is found that the *virial theorem* can be often applied locally in such zones and that in a loge the correlation between the positions of electrons is strong but it is weak between the electrons moving in distinct loges. Finally, chemical bonds appear to be regions possessing a certain autonomy and it is proved that their organisation effectively resists strong perturbations. Thus the bond loge of BeH remains almost unchanged in HBeH. The addition of a new atom has perturbed the BeH loge very little. It is practically transferable. One can thus consider the problem of building molecules with a loge-meccano.

4. SURVEY ON THE APPLICATIONS OF THEORETICAL MOLECULAR PHYSICS

In such a short account, one cannot analyse the numerous applications of the setting up of a mathematical government among molecular populations. With the help of the theory of collisions and even without any empirical parameter, one can calculate the cross-sections of chemical reactions which physicists produce in *crossed molecular beams experiments*. The experimental agreement is excellent.

Quantum mechanics is applied to understand the nature of the molecules necessary for the origin of life on the earth, to analyse the mutation mechanism, to understand the retinal effect on the vision mechanism, to guide the synthesis of new antibiotics and to the understanding of chemical cancer etiology [5].

Centre de Mécanique Ondulatoire Appliquée, CNRS, Paris

NOTE

[1] In a large number of scientific considerations, there exist almost always some contradictions.

BIBLIOGRAPHY

[1] Heidegger, M., *L'être et le temps*, Gallimard (1964).
[2] Popper, K., *La logique de la découverte scientifique*, Payot (1973).
[3] Kolos, W. and Roothaan, C. C. J., *Rev. Mod. Phys.* 32 (1960), 219.
[4] For a more technical account and the bibliography, see Daudel, R., in *Wave Mechanics, The First Fifty Years*, Butterworth (1973) p. 61; Daudel, R., Bader, R. F. W., Stephens, M. E., and Borrett, D. S., *Can. J. Chemistry* 52 (1974), 1310.
[5] For instance, see Daudel, R., *La Chimie Quantique*, Que sais-je? Presses Universitaires de France (1973).

GEORGES LOCHAK

HIDDEN PARAMETERS, HIDDEN PROBABILITIES*

ABSTRACT. In this paper is analysed the proof given by J. S. Bell of an inequality between mean values of measurement results which, according to him, would be characteristic of any local hidden-parameter theory. It is shown that Bell's proof is based upon a hypothesis which was already present in von Neumann's famous theorem: it consists of admitting that hidden values of parameters must obey the same statistical laws as observed values. This hypothesis contradicts in advance well-known and certainly correct statistical relations in measurement results: one must therefore reject the type of theory considered by Bell and his inequality has no general meaning.

Since the beginning of Wave Mechanics, various authors have tried to prove that it is a purely statistical theory, the probabilistic laws of which would be incompatible with the existence of a subjacent determinism. The most famous theorem which attempted to prove this result was given by von Neumann[1], but it was later refuted by Louis de Broglie, who was guided by his analysis of the statistical pattern of Wave Mechanics and by a description of the process of measurement in the framework of double-solution theory (de Broglie[2, 3]). Other arguments have been given later against this theorem in [4] and [5].

Although the idea of a 'purely statistical theory' has still many adepts, it seems that no one any longer defends the von Neumann theorem. In recent years, however, a theorem of Bell[6, 7] has been frequently cited, by which any local hidden-variable theory would necessarily entail a certain inequality between expected values in a measurement process. As this inequality, according to Bell, is true for any local hidden-variable theory and false for the orthodox theory, one tends to take it as a criterion in the choice between the two types of theories, and as a recent experiment[8, 9] seems to contradict this inequality, it might appear as a condemnation of all hidden-variable theories. Bell's hypothesis of 'locality' amounts to saying that it is always possible to move one of the two particles sufficiently far away from the other so that a measurement performed on one will not affect a measurement performed on the other. In our opinion, this hypothesis is well-founded, but we must specify that it involves a

J. Leite Lopes and M. Paty (eds.), Quantum Mechanics, a Half Century Later, 245–259. All Rights Reserved
Copyright © 1976 by the Foundations of Physics journal

finite extension of the wave-packets associated with the particles. This implication was pointed out long ago by Louis de Broglie[10, 2], and he has recently developed it[11], but it seems to have been neglected both by Bell and by the other authors who wrote about his theorem. Now, it is an essential point since it introduces doubt about the use by Bell of the quantum formula of singlet state in the description of correlated measures. However, we will leave this question aside in the present paper, for we would like to examine another aspect of Bell's reasoning: the statistical distributions of hidden variables.

Our intention herein is to show that Bell's reasoning involves not only a hypothesis on the local character of the theory, but also a statistical hypothesis which is exactly the one introduced by von Neumann in his famous theorem and which is more or less clearly supposed by most of the authors who reflect on hidden variable theories. This hypothesis consists in admitting that such a theory must restore the classical probabilistic pattern (we shall define it later) simultaneously in the statistics of all *measurement results*. But it leads immediately to a contradiction with the calculation of the mean values in wave mechanics since the latter violates the usual pro-babilistic pattern: therefore it is not astonishing to 'discover' after-wards an incompatibility between the results of wave mechanics and those ascribed to hidden variable theories.

On the contrary, we shall try to prove by means of the example of the measurement of a spin component, that the Theory of double-solution of Louis de Broglie does not contain this defect and does not fall under the framework of theories considered by Bell. But let us first recall the substance of Bell's reasoning.

Let us denote by a and b two particles with spin 1/2 initially carried by the same wave packet and being in a certain quantum state which we suppose entirely described by a set of hidden parameters collectively denoted by λ, with a probability density $\rho(\lambda)$.

Suppose then that, these two particles having moved away one from the other, their spins are measured separately in two directions, respectively labeled **a** and **b**. The result of the first measurement, Bell says, will be $A = \pm 1$ and can depend on the parameters λ and on the direction **a**, but not in the direction **b**, for he supposes – here lies his hypothesis of 'locality' – that the two particles are separated at the instant of the measurement. In the same way, he says, the measurement of the spin of b will give us $B = \pm 1$ and can depend on λ and in the direction $\mathbf{b}$ but not in the direction $\mathbf{a}$.

Writing now $P(\mathbf{a}, \mathbf{b})$ for the mean value of the product AB Bell gives the definition:

(1) $P(\mathbf{a}, \mathbf{b}) = \int d\lambda\, \rho(\lambda) A(\mathbf{a}, \lambda) B(\mathbf{b}, \lambda).$

We can certainly agree with such a formula but the problem will be: can we introduce the same $\rho(\lambda)$ for all the mean values? We shall see later that it is possible for hidden values but not always for measured values. Bell, then, introduces two other directions $\mathbf{a}'$ and $\mathbf{b}'$, which correspond to another setting up of the measurement instruments, and he studies the difference:

(2) $P(\mathbf{a}, \mathbf{b}) - P(\mathbf{a}, \mathbf{b}') = \int d\lambda\, \rho(\lambda)[A(\mathbf{a}, \lambda)B(\mathbf{b}, \lambda)$

$- A(\mathbf{a}, \lambda)B(\mathbf{b}', \lambda)].$

By adding and subtracting the same expression in the integral, he writes this equality as:

(3) $P(\mathbf{a}, \mathbf{b}) - P(\mathbf{a}, \mathbf{b}') = \int d\lambda\, \rho(\lambda)A(\mathbf{a}, \lambda)B(\mathbf{b}, \lambda)$

$\times [1 + A(\mathbf{a}', \lambda)B(\mathbf{b}', \lambda)]$

$- \int d\lambda\, \rho(\lambda)A(\mathbf{a}, \lambda)B(\mathbf{b}', \lambda)$

$\times [1 \pm A(\mathbf{a}', \lambda)B(\mathbf{b}, \lambda)].$

Now, in one way, we know that $\rho(\lambda)$ is normalized, and in the other way, we have $|AB| = 1$. The result of this last equality is that the two expressions in square brackets under the integral signs are positive and:

(4) $|P(\mathbf{a}, \mathbf{b}) - P(\mathbf{a}, \mathbf{b}')| \leqslant \int d\lambda\, \rho(\lambda)[1 + A(\mathbf{a}', \lambda)B(\mathbf{b}', \lambda)]$

$+ \int d\lambda\, \rho(\lambda)[1 + A(\mathbf{a}', \lambda)B(\mathbf{b}, \lambda).$

It follows, by definition (1):

$|P(\mathbf{a}, \mathbf{b}) - P(\mathbf{a}, \mathbf{b}')| \leqslant 2 \pm [P(\mathbf{a}', \mathbf{b}') + P(\mathbf{a}', \mathbf{b})]$

and finally:

(5) $|P(\mathbf{a}, \mathbf{b}) - P(\mathbf{a}, \mathbf{b}')| + |P(\mathbf{a}', \mathbf{b}') + P(\mathbf{a}', \mathbf{b})| \leqslant 2.$

This is Bell's inequality in the form given by Clauser *et al.*[8] as they proposed at the same time an experimental test which was later performed by Freedman and Clauser and the result of which was negative[9].

Let us now inspect the hypotheses on which Bell's reasoning is based and which are included in the definition of the mean values.

One of these is the hypothesis of locality which we will not further study in this paper.

But a second hypothesis (of a statistical nature) is given by Bell as self-evident, while according to us, it vitiates his reasoning: it consists in calculating the mean values of the measurement results with the help of the probability density $\rho(\lambda)$ of the hidden parameters in the *initial state* of the system. The statistics so introduced on the measurement results do not depend on the measurement process itself. Clauser *et al.* even insist on this point and say: "... since the pair of particles is generally emitted by a source in a manner physically independent of the adjustable parameters **a** and **b**, we assume that the normalized probability distribution $\rho(\lambda)$ characterizing the ensemble is independent of **a** and **b**". This remark is certainly correct if applied to the *initial* probability distribution of the hidden variables but that is not the question; the question is: have we to use just this probability distribution for the calculation of mean values of measurement results? Let us see now some immediate consequences of Bell's hypothesis.

We have, according to this hypothesis, a *single* probability distribution for the calculation of the mean values of *all* the measurement results. With the help of this distribution $\rho(\lambda)$ we can easily define probability $\mathscr{P}_a(\alpha)$ to find the value $A(\mathbf{a}, \lambda) = \alpha$ ($\alpha = \pm 1$) as a result of the measurement of the spin of the particle α in the direction **a**; and we can find a similar probability related to the measurement of the spin of the same particle in another direction **a**'.

Indeed, if we define in the configuration space $\mathscr{E}\{\lambda\}$ of the hidden parameters the two following subspaces:

$$(6) \quad \begin{cases} \mathscr{E}_\alpha = \mathscr{E}_{A(\mathbf{a},\lambda)=\alpha} = \mathscr{E}\{\lambda/A(\mathbf{a}, \lambda) = \alpha\} & (\alpha = \pm 1), \\ \mathscr{E}_{\alpha'} = \mathscr{E}_{A(\mathbf{a}',\lambda)=\alpha'} = \mathscr{E}\{\lambda/A(\mathbf{a}', \lambda) = \alpha'\} & (\alpha' = \pm 1), \end{cases}$$

we obtain the probabilities:

$$\mathscr{P}_a(\alpha) = \mathrm{Pr}\{A(\mathbf{a}, \lambda) = \alpha\} = \int_{\mathscr{E}_\alpha} \rho(\lambda) \, d\lambda,$$

$$(7) \qquad \mathscr{P}_{a'}(\alpha') = \Pr\{A(a', \lambda) = \alpha'\} = \int_{\mathscr{E}_{\alpha'}} \rho(\lambda)\, d\lambda.$$

Now, we shall also be able to define the quantity:

$$(8) \qquad \mathscr{P}_{a,a'}(\alpha, \alpha') = \Pr\{A(a, \lambda) = \alpha, A(a', \lambda) = \alpha'\} = \int_{\mathscr{E}_{\alpha} \cap \mathscr{E}_{\alpha'}} \rho(\lambda)\, d\lambda$$

and this is nothing but the probability of finding value α for the spin measure of particle α in the direction a and value α' for the spin measure of the same particle in another direction a'.

As a measurement of a spin can only give the values ± 1 (reduced here to unity) we shall have:

$$(9) \qquad \mathscr{E}_{\alpha=-1} \cup \mathscr{E}_{\alpha=1} = \mathscr{E}\{\lambda\}; \; \mathscr{E}_{\alpha'=-1} \cup \mathscr{E}_{\alpha'=1} = \mathscr{E}\{\lambda\}.$$

Therefore:

$$(10) \qquad \int_{\mathscr{E}_{\alpha} \cap \mathscr{E}_{\alpha'=-1}} \rho(\lambda)\, d\lambda + \int_{\mathscr{E}_{\alpha} \cap \mathscr{E}_{\alpha'=1}} \rho(\lambda)\, d\lambda = \int_{\mathscr{E}_{\alpha}} \rho(\lambda)\, d\lambda,$$

whence, according to Definitions (7) and (8), we deduce:

$$(11) \qquad \mathscr{P}_{a}(\alpha) = \sum_{\alpha'=\pm 1} \mathscr{P}_{a,a'}(\alpha, \alpha'); \; \mathscr{P}_{a'}(\alpha') = \sum_{\alpha=\pm 1} \mathscr{P}_{a,a'}(\alpha, \alpha'),$$

which is the theorem of total probabilities.

Let us now define two new probabilities:

$$(12) \qquad \mathscr{P}_{a}^{(a')}(\alpha, \alpha') = \frac{\mathscr{P}_{a,a'}(\alpha, \alpha')}{\mathscr{P}_{a'}(\alpha')}; \; \mathscr{P}_{a'}^{(a)}(\alpha, \alpha') = \frac{\mathscr{P}_{a,a'}(\alpha, \alpha')}{\mathscr{P}_{a}(\alpha)}.$$

The first is the probability of obtaining the value α as a result of the measurement of the projection on a of the spin of the particle a when we already know that the a' component equals α'. The second is the probability of finding the value α' for the a' component if the value of the a component is already known.

According to (11) we shall have:

$$(13) \qquad \mathscr{P}_{a}(\alpha) = \sum_{\alpha'=\pm 1} \mathscr{P}_{a}^{(a')}(\alpha, \alpha') \mathscr{P}_{a'}(\alpha');$$

$$\mathscr{P}_{a'}(\alpha) = \sum_{\alpha=\pm 1} \mathscr{P}_{a'}^{(a)}(\alpha, \alpha') \mathscr{P}_{a}(\alpha),$$

and therefore we find the theorem of compound probabilities.

The relations (11), (12), (13) constitute what we call the 'classical pattern of statisticians'. They reside in the existence of a probability $\mathscr{P}_{\mathbf{a},\mathbf{a}'}(\alpha, \alpha')$ and consequently in the implicit hypothesis that it is possible to find, via a single measuring instrument or two mutually compatible instruments, the two values α and α' for the projection of the spin of a on $\mathbf{a}$ and $\mathbf{a}'$.

Before discussing in greater detail the problem of the statistical pattern of a hidden-variable theory, let us note that the projections of the spin of a particle on two different directions are not simultaneously measurable, and that is why the probability $\mathscr{P}_{\mathbf{a},\mathbf{a}'}(\alpha, \alpha')$ does not exist in the usual statistical pattern of wave mechanics. Therefore Bell's hypothesis goes against usual calculations concerning measurement results; in particular, it goes against the Heisenberg relations. Thus, it is not surprising that Bell draws conclusions from his hypothesis which are in contradiction with usual wave mechanics; in the entire experimental field in which this theory has meaning, Bell's conclusions will be found necessarily to be false.

It must be pointed out that Bell effectively makes use of his statistical hypothesis in the proof of his theorem. He uses it in the formula (2), by which is expressed the difference $P(\mathbf{a}, \mathbf{b}) - P(\mathbf{a}, \mathbf{b}')$ and where the two mean values are deduced from the single density $\rho(\lambda)$ though one of the averages concerns the measure of spin of particle b in the direction $\mathbf{b}$, while the other contains the result of a measurement of the spin of the *same* particle in *another* direction $\mathbf{b}'$. Yet in this formula we shall see further that in reality two different densities $\rho(\lambda, \mathbf{b})$ and $\rho(\lambda, \mathbf{b}')$ should be introduced. But the implicit hypothesis of Bell appears more clearly when, to pass from the expression (2) to the mathematically equivalent expression (3), we add and subtract under the sign of integration the product of functions:

$$\pm\rho(\lambda)A(\mathbf{a}, \lambda)A(\mathbf{a}', \lambda)B(\mathbf{b}, \lambda)B(\mathbf{b}', \lambda),$$

which supposes that it is possible to know by a measurement process, performed in the same state λ of the system, the components $\mathbf{a}$ and $\mathbf{a}'$, of the spin of particle a and the components $\mathbf{b}$ and $\mathbf{b}'$ of the spin of particle b. One could object to this: "It is true that we cannot measure in one experiment two different projections of the spin of b in the directions $\mathbf{b}$ and $\mathbf{b}'$ but we can conceive, *for the same state λ of this particle*, two different experiments for the measure of each of these projections and define the probability $\mathcal{P}_{b,b'}(\beta, \beta')$ to find β if we measure the projection **b** and β' if we measure the projection **b'**."

This objection is false because the impossibility of a simultaneous measure of two different spin-projections of the same particle is not due to a simple incompatibility of instruments: it comes from the fact that *the state in which we must put the particle to measure its spin-component* **b** *is not the same as the one in which we must put it to measure the component* **b'**.

Therefore, the measurement process of a spin-projection, by changing the state of the particle, changes also the statistical predictions about the measurement of another projection and it is what forbids to define in a given state of a particle a density of probability such as $\mathcal{P}_{a,a'}(\alpha, \alpha')$ or $\mathcal{P}_{b,b'}(\beta, \beta')$.

Thus (this problem is widely analysed in [2] and [12]) we know that, for two non-simultaneously observable quantities (like spin components) we have

$$\mathcal{P}_{a'}(\alpha')\mathcal{P}_a^{(a')}(\alpha, \alpha') \neq \mathcal{P}_a(\alpha)\mathcal{P}_{a'}^{(a)}(\alpha, \alpha').$$

In other words, the probability of finding the values α and α' by measuring two spin-components **a** and **a'** depends on the order of the measurements. But this fact and so the non-existence of the probability $\mathcal{P}_{a,a'}(\alpha, \alpha')$ is not a 'taint' of the theory, it is a consequence of the wave-particle dualism: if a hidden-parameter theory contradicts this fact, it will necessarily go against a lot of correct results of usual wave mechanics.

A fundamental point seems to be the following: when a hidden-variable theory introduces quantities which are simultaneously defined and are governed by a classical statistical pattern, it follows from this:

> neither that the measured values of these quantities are the ones they had before the measurement;
> nor that the classical statistical pattern is observable.

When we wish to explain the statistical laws of wave mechanics by supposing the existence of a subjacent determinism, we are led to consider that each particle, *before any observation*, is in a well-defined state, where all the physical quantities, which characterize the particle, are *simultaneously defined* by a set of parameters λ which are hidden because we do not know their exact values; these values will be only characterized by a probability distribution $\rho(\lambda)$. Up to now, we agree with Bell, and also with Clauser *et al.*, when they say that there is obviously no reason to suppose any dependence of this distribution on any future measurement.

Let us now consider a certain direction **a** in the space. The spin of the particle has, according to our hypothesis and *before* any measurement, a certain projection on **a**, which depends on parameters λ and which we call $s(\mathbf{a}, \lambda)$; we have, similarly, a projection $s(\mathbf{a}', \lambda)$ on another direction **a**'. But, just as the parameters λ on which they depend, quantities $s(\mathbf{a}, \lambda)$ and $s(\mathbf{a}', \lambda)$ are *hidden quantities* and they must not be mistaken for the result of the measurement of these projections denoted by Bell: $A(\mathbf{a}, \lambda)$ and $A(\mathbf{a}', \lambda)$. In particular, these quantities s will not assume, generally, and certainly not both at the same time, the quantized values obtained by a measurement: their values α and α' can fill here a certain continuum.

On these hidden quantities $s(\mathbf{a}, \lambda)$ and $s(\mathbf{a}', \lambda)$, we can now build statistics and they will follow the traditional pattern since all the quantities are simultaneously *defined* (but attention! We do not say 'simultaneously *measurable*'). To simplify we shall write s and s' for the above quantities and we shall denote Λ, S, S', as random variables whose possible values are respectively λ, s and s'. First, we obtain the densities of conditional probabilities

(14)
$$\begin{cases} \rho_S^{(\Lambda)}(\alpha, \lambda) = \delta(\alpha - s(\lambda)), \\ \rho_{S'}^{(\Lambda)}(\alpha', \lambda) = \delta(\alpha' - s'(\lambda)), \end{cases}$$

which simply mean that, if Λ takes the value λ, S takes the value $\alpha = s(\lambda)$ and S' the value $\alpha' = s'(\lambda)$. Then, the probability densities for S and S' will be:

(15)
$$\begin{cases} \rho_S(\alpha) = \int \rho(\lambda)\delta(\alpha - s(\lambda))\, d\lambda, \\ \rho_{S'}(\alpha') = \int \rho(\lambda)\delta(\alpha' - s'(\lambda))\, d\lambda, \end{cases}$$

and the probability that S takes a value included in the interval $[\alpha, \alpha + d\alpha]$ *and* S' a value included in the interval $[\alpha', \alpha' + d\alpha']$ will be defined by a density:

(16)
$$\rho(\alpha, \alpha') = \int \rho(\lambda)\delta(\alpha - s(\lambda))\delta(\alpha' - s'(\lambda))\, d\lambda.$$

One can verify that:

$$(17) \qquad \rho_S(\alpha) = \int \rho(\alpha, \alpha') \, d\alpha'; \; \rho_{S'}(\alpha') = \int \rho(\alpha, \alpha') \, d\alpha.$$

So, we have a conventional statistical pattern; we can easily find the averages of S and S' and, provided that the values taken by these quantities, and therefore the values of the functions $s(\lambda)$ and $s'(\lambda)$, remain limited, we may deduce inequalities of the same kind as those given by Bell.

But is this statistical pattern an observable one? Can we verify it by a measurement process? This is exactly what Bell assumes in his theorem (and it was already assumed in the theorem of von Neumann): Bell assumes that the values taken by hidden quantities such as $s(\lambda)$ and $s'(\lambda)$, and the probability distributions of these values are those which we can observe in an experiment. This error leads Bell to the same truism as von Neumann, that is to say that traditional pattern is incompatible with statistics of measurement results in usual wave mechanics. On the contrary, we absolutely must assume that measurement process modifies, in general, the state of the observed system, so that not only the measured values of the physical quantities are not necessarily those taken by the hidden quantities before the measurement, but also the probabilities we have just defined are generally not the observed ones: they are *hidden probabilities*.

So, it was perfectly licit and even evident to suppose that the initial probability distribution $\rho(\lambda)$ does not depend on **a** (or on **b**) but this distribution is not, and cannot be, the one that we need for the statistics of measurement results: we can be sure of this assertion, precisely because, if we adopt this initial density $\rho(\lambda)$, we obtain a traditional statistical pattern on measurement results, which obviously contradicts the well-known and certainly true statistical results in wave mechanics.

The problem is, therefore, to understand, by analysing the process of measurement, how the apparatus modifies the state of the system so as to reveal the observed values and the probability distributions correctly calculated by usual wave mechanics. Just such an analysis has been performed by Louis de Broglie [2] in the framework of double solution theory by describing not a measurement of spin projections, but the measurement of impulse and position of a particle, which is another typical example of two quantities which are not simultaneously measurable. We shall now briefly expose here these considerations by taking the example of a measurement of a spin component.

The theory of measurement in the theory of double solution [2, 13, 14] rests on two fundamental ideas. The first one is that the measurement of any physical quantity attached to a particle ultimately comes to a *measurement of position*. In the so-called measurements of 'first kind' the role of the measurement apparatus consists in modifying the state of the motion of the observed corpuscle in such a way that the corpuscle is compelled to choose between several possible directions and that, in each one of them, the physical quantity one wants to measure has a certain fixed value which is but one of the eigenvalues of the corresponding quantum operator. Thus, if the corpuscle is found to be in a certain direction at the outlet of the apparatus one is able to attach to it unambiguously one of the eigenvalues of the physical quantity in question. This is the way one obtains the orientation of the spin in a Stern-Gerlach experiment or measures its amplitude in the corpuscular jet apparatus of Rabi. In fact, in this later case, in order to perform the measurement, one must search a Hertzian resonance by means of the 'outsignal' of the jet, that is to say, by measuring the probability of presence of the corpuscles in a certain direction when they get out of the apparatus.

We shall simply mention here the measurements of 'second kind' (which shall not be dealt with here) in which the particle whose position is detected is not the one which is studied but a secondary particle instead arising from a collision with the first one or spontaneously emitted by it. Actually, this case corresponds to the experiment studied by Bell (experiment of the Einstein, Podolski and Rosen type) but the simpler case of measurements of 'first kind' will be enough for our purposes since we intend to study the statistical consequences – which, according to us, are false – brought about by Bell's hypothesis for *each one* of the two particles a and b considered independently of the other.

The second important idea, which is but a consequence of the first one, is that if one wants to measure a physical quantity by registering the position of the corpuscle, the measurement apparatus must split the initial wave – which embodies the corpuscle – in several wave trains separated in the space, each one of which corresponds to a fixed value of the physical quantity in such a way that, if the corpuscle is found to be in one of such wave trains, one may then infer unambiguously from this the value one looked for. One must point out that the polarisers placed at the outlet and the inlet of the Rabi's apparatus above-mentioned, aim precisely at that separation of the wave trains, and the Stern-Gerlach apparatus is but a polariser itself.

Finally one must recall that the two preceding ideas are tied to the fundamental idea of the theory of the double solution which is the permanent localisation of the corpuscles. We now want to summarize a few essential points of this conception which has been developed with greater detail in earlier publications[15]. We shall assume that the corpuscle in its motion is guided by a certain wave v to which it is tied, which has a weak amplitude and is unable to stimulate the registering apparatuses known at the present. This wave is a solution of the Schrödinger equation (or of some other equation of wave mechanics) excepting in a small region around the corpuscle and perhaps over the edges of the wave trains.

The usual wave ψ, which is an element of probabilistic prediction and thus carries a somewhat subjective character by opposition to the material character of wave v, will be represented by a function proportional to v ($\psi = cv$) and normalized to unity. The probability of presence of the corpuscle in a point of the space will be $|\psi|^2$.

Let us now assume that Q is the physical quantity in question, and denote by the same letter the corresponding quantum operator, by q_k its eigenvalues and by φ_k its eigenfunctions which we take as being normalized to unity. Since these functions span the whole space, we may then write the following for the wave v:

$$(18) \qquad v = \sum_k \gamma_k \varphi_k \qquad (\gamma_k = \text{const.}),$$

and similarly for the function ψ:

$$(19) \qquad \psi = \sum_k c_k \varphi_k \qquad (c_k = c\gamma_k),$$

where c is the proportionality constant between v and ψ (normalization constant).

The measurement apparatus will then parcel out the initial wave v in several distinct wave trains, each one of which corresponds to one of the components of development (18).

According to the theory of the double solution, the particle was localized in wave v before entering the measurement apparatus and, after having followed a usually very complicated path, it will be found when getting out the apparatus, in one of the wave trains φ_k. If we knew the initial conditions, i.e., the initial localization of the particle in the wave, we could predict precisely its wave train at the outlet by calculating its path, and therefore its eigenvalue q_k. But since we know but the position probability density in the initial wave, $\|\psi(\mathbf{r})\|$, we can just predict the probabilities $|c_k|^2$ of finding the particle in one wave train φ_k and thus attach to it a value q_k.

Let us now go back to the measurement of a spin component. For the sake of simplicity, we shall consider the non-relativistic case so that the wave associated to the particle is represented by a spinor with two components obeying the Pauli equation. The hidden parameters which Bell denoted by λ are now the position coordinates $\mathbf{r}$ of the particle in the wave; therefore the density $\varphi(\lambda)$ will be written

$$(20) \qquad \rho(\mathbf{r}) = \psi^*(\mathbf{r})\psi(\mathbf{r}) = |\psi_1(\mathbf{r})|^2 + |\psi_2(\mathbf{r})|^2,$$

with

$$(21) \qquad \psi = \begin{pmatrix} \psi_1 \\ \psi_2 \end{pmatrix}, \qquad \int \rho(\mathbf{r})\,dv = 1.$$

In any state ψ of the particle, we shall define a spin[16]

$$(22) \qquad s(\mathbf{r}) = \frac{\psi^* \sigma_{op} \psi}{\psi^* \psi},$$

where σ_{op} denotes the set of the three Pauli matrices:

$$(23) \qquad \sigma_1 = \begin{pmatrix} 0 & 1 \\ 1 & 0 \end{pmatrix}; \qquad \sigma_2 = \begin{pmatrix} 0 & -i \\ i & 0 \end{pmatrix}; \qquad \sigma_3 = \begin{pmatrix} 1 & 0 \\ 0 & -1 \end{pmatrix}.$$

Formula (22) means that if the particle is found at point $\mathbf{r}$ of the wave $\psi(\mathbf{r})$, it then has the spin $s(\mathbf{r})$. But, for the time being, the position $\mathbf{r}$ of the particle remains unknown to us, and we know but the probability density $\rho(\mathbf{r})$ given by (20); *the spin is thus a hidden variable* and we may write for it the formulae (14) to (17) by simply replacing $\rho(\mathbf{r})$ and $s(\mathbf{r})$ by their expressions (20) and (22). But it is important to remark that the probabilities defined by these formulae (14) to (17) are themselves hidden too, for we can not measure the spin components without disturbing the form of the wave $\psi(\mathbf{r})$ and therefore without changing the density and thus all the other probability distributions.

For us to be sure let us first consider the measurement of a spin component parallel to a certain direction $\mathbf{a}$ of the space. This can be done by letting the particle entering in an inhomogeneous magnetic field oriented along direction $\mathbf{a}$. This field will split the wave in two distinct wave trains which shall be such that if the particle is found to be in one of them we may then be sure that $s(\mathbf{a}, \mathbf{r}) = 1$ and if it is found in the other then we know that $s(\mathbf{a}, \mathbf{r}) = -1$.

Let us then take for direction $\mathbf{a}$ the z axis of the space. We shall have, according to (22) and (23).

$$(24) \qquad s(\mathbf{a}, \mathbf{r}) = s_3(\mathbf{r}) = \frac{\psi^* \sigma_3 \psi}{\psi^* \psi},$$

and so,

$$(25) \qquad s_3(\mathbf{r}) = \frac{\psi_1^* \psi_1 - \psi_2^* \psi_2}{\psi_1^* \psi_1 + \psi_2^* \psi_2}.$$

The wave train for which $s_3 = 1$ will then be such that

$$(26) \qquad \psi^+ = \begin{pmatrix} \psi_1 \\ 0 \end{pmatrix},$$

and the one for which $s_3 = -1$ will be

$$(27) \qquad \psi^- = \begin{pmatrix} 0 \\ \psi_2 \end{pmatrix}.$$

The total wave at the outlet of the magnetic field will be $\psi = \psi^+ + \psi^-$ but since the two components are separated in space, filling two distinct regions R^+ and R^-, the probability of getting the value $+1$ of the component s_3 of the spin will be

$$(28) \qquad \mathscr{P}_3(+) = \int_{R^+} \psi^* \psi \, dv = \int |\psi_1|^2 \, dv,$$

while the probability of getting the value -1 will be written

$$(29) \qquad \mathscr{P}_3(-) = \int_{R^-} \psi^* \psi \, dv = \int |\psi_2|^2 \, dv.$$

The normalization of the wave assures us that

(30) $\mathscr{P}_3(+) + \mathscr{P}_3(-) = 1.$

It is understood that only the s_3 component of the spin has its values distributed according to the elementary probability law given in $\mathscr{P}_3(+)$ and $\mathscr{P}_3(-)$ and one also sees that such law is different from law (15) that yielded the hidden distributions before the measurement.

Let us now assume that we wanted to measure the spin component along another direction distinct from **a**; let us denote it by **a'**. We should then set up another apparatus in all points similar to the one above, except that its magnetic field would then be orientated along direction **a'**, which would now split again the initial wave in two wave trains (*obviously different from those we had obtained above*). Each one of these wave trains would correspond to one of the values +1 or −1 of the component $s(\mathbf{a'}, \mathbf{r})$. We could then get, for these two values, certain probabilities $\mathscr{P}'_3(+)$ and $\mathscr{P}'_3(-)$ with expressions similar to those we considered above; but in order to do this, it is necessary to change the reference frame and take for the z axis no longer the direction **a** but rather the direction **a'** and in Equations (28) and (29), we obviously have now new functions ψ'_1 and ψ'_2.

It is thus seen that if one assumes (as it is done in the theory of double solution) that the hidden parameters are the coordinates of the particle, the measurement of two different components of a spin requires two partitions of the initial wave train that are different and incompatible with each other. It clearly follows that the probability density $\rho(\mathbf{r})$ which was true for the initial wave, will give rise, when arising from the magnetic field, to two different densities $\rho(\mathbf{a}, \mathbf{r})$ and $\rho(\mathbf{a'}, \mathbf{r})$ corresponding to the measurements of component $s(\mathbf{a})$ and component $s(\mathbf{a'})$ of the spin. This is what forbids us to write a formula similar to (2) which, in fact, is only possible for the hidden distributions and not for the results of measurements.

Let us point out, to conclude, that the foregoing developments aim only to refute the *reasoning* according to which Bell got his inequality.

But this refutation does not imply that the inequality in question is not satisfied in certain cases of measurements of correlated events. In fact, Bell's inequality will be true whenever the correlation between two measurements will be weakened enough (as it is seen from formula (5)) *independently of the cause of such weakening*. One may then say, in general, that if for some reason the initial state of the system of the two particles (or an intermediary state) is not a pure case but a mixture, the inequality has some chances of being satisfied. Such circumstances might arise, for instance, in the emission of two photons in cascade if the intermediary level is degenerate and has a lifetime long enough to let a mixture to be produced between the degenerate states.

But this means that one must be extremely cautious with regard to the 'crucial' character one could be tempted to attach to such experiments of correlation *whatever their result may be.*

Fondation Louis de Broglie,
Paris

NOTE

* Printed in *Foundations of Physics* 6 (1976), 173, with some modifications, under the title: 'Has Bell's Inequality a General Meaning for Hidden-Variable Theories?'

BIBLIOGRAPHY

[1] von Neumann, J., *Mathematical Foundations of Quantum Mechanics*, Princeton University Press, Princeton, 1955.
[2] de Broglie, L., *La théorie de la mesure en mécanique ondulatoire (interprétation usuelle et interprétation causale)*, Gauthier-Villars, Paris, 1957.
[3] de Broglie, L., *Étude critique des bases de l'interprétation actuelle de la mécanique ondulatoire*, Gauthier-Villars, Paris, 1963 (English transl., Elsevier, Amsterdam).
[4] Mugur Schächter, M., *Étude du caractère complet de la théorie quantique*, Gauthier-Villars, Paris, 1964.
[5] Bell, J. S., *Rev. Mod. Phys.* 38 (1966), 447.
[6] Bell, J. S., *Physics* 1 (1964), 195.
[7] Bell, J. S., *Foundations of Quantum Mechanics*, Proc. Int. School of Physics 'Enrico Fermi', Varenna, 1971.
[8] Clauser, J. F., Horne, M. A., Shimony, A., and Holt, R. A., *Phys. Rev. Lett.* 23 (1969), 880.
[9] Freedman, S. J. and Clauser, J. F., *Phys. Rev. Lett.* 28 (1972), 938.
[10] de Broglie, L., *La Mécanique ondulatoire des systèmes de corpuscules*, Gauthier-Villars, Paris, 1939 (reedited in 1950).
[11] de Broglie L., *Comptes rendus Acad. Sci.* 278B (1974), 721.
[12] de Broglie, L., *La Revue Scientifique* No. 3292, fasc. 5, 87 (1948), 259.
[13] Andrade e Silva, J., *Foundations of Physics* 2 (1972), 245.
[14] Andrade e Silva, J., *Foundations of Quantum Mechanics*, Proc. Int. School of Physics 'Enrico Fermi', Varenna 1971.
[15] de Broglie, L., *La réinterprétation de la Mécanique ondulatoire*, Gauthier-Villars, Paris, 1971.
[16] de Broglie, L., *Comptes rendus Acad. Sci.* 272B (1971), 349.

MICHEL PATY*

THE RECENT ATTEMPTS TO VERIFY QUANTUM MECHANICS

What is needed now is a hypothetical tentative approach, to attempt both by theory and by experiment to inquire into the conditions in which quantum mechanics might break down, to reveal a new structure of physical law and a new order in physical movement. Experiments devised in order to study questions raised in such an inquiry could, in principle, falsify the basic principles of quantum mechanics and show the need for new ones.

D. Bohm, J. Bub, *Rev. Mod. Phys.* **38** (1966), 469.

ABSTRACT. We attempt to give an account of the experiments which deal with the verification of quantum mechanics and the hidden variables problem. First, we recall the well-known EPR paradox which, in spite of its refutation by Bohr, was the starting point of the questioning on the completeness of quantum mechanics and of hidden variable theories; we then recall Bell's theorem, which shows that the two approaches, quantum mechanics and hidden variables, can be put in contradiction. We describe thereafter the various types of experiments which have been carried out on that subject, mostly concerning the correlation measurements between two photons emitted by a quantum system. The most recent experimental results are contradictory, some of them appearing to confirm and others to contradict quantum mechanics.

A review of these is given and a discussion is presented about their possible implications.

1. INTRODUCTION

For the last fifty years or so, quantum mechanics has had a well established position; strong in its internal coherence and its multiple successes in various spheres of physics, it appears as an extremely powerful theory whose perfection and principles are not questioned. No experiment (amongst the many which have been performed since its birth) has yet contradicted it (at least up to now as we shall see further on) and a theory which would claim to be more refined might consequently seem unnecessary. However, the problems raised by its

* Postal address: 67037 Strasbourg-Cedex, France.

J. Leite Lopes and M. Paty (eds.), Quantum Mechanics, a Half Century Later, 261–289. All Rights Reserved

interpretation have given rise to alternative hypotheses amongst which the hidden variables hypothesis is the best known and has the most evident meaning. Its requirements logically follow the line of the objections raised by Einstein, Podolsky and Rosen. I will therefore begin this account by recalling them, and then sketch briefly the path which has led from a hidden variable theory compatible with the statistical determination of quantum mechanics to the recent demonstration of the incompatibility in some cases of these two representations (Bell's theorem). At the same time, experiments were performed and interpreted as a function of the theoretical developments. Periodically repeated and refined, they had, up to recently, confirmed the predictions of quantum mechanics. Some recent results seem to throw doubt on this fine certainty. In fact, taken together, the results are contradictory; further experiments have already been reported at various places. Such is the paradox of this mechanics, which is so powerful in the analysis of infinitely small structures of the matter by means of the most 'sophisticated' technology, that it can lend itself to the most radical reexamination in simple experiments.

2. EPR PARADOX AND BELL'S THEOREM

2.1. *EPR Paradox*

The objection raised by Einstein, Podolsky and Rosen[1–6] in 1935, and known as EPR paradox, is based on the assertion of specific realism. According to this, each *element of physical reality* is represented by a corresponding quantity in a complete physical theory. The *sufficient* criterion for representing this physical reality is stated as follows: if the value of a physical quantity can be predicted with certainty, without perturbing the system, then there exists an element of physical reality corresponding to this quantity. From this basis, the authors wish to prove that quantum mechanics is insufficient, by showing that it cannot offer a *complete* description of all the physical factors or *elements* of reality of a system under consideration. Let us note at once that these assertions contain implicit hypotheses concerning physical reality, hypotheses which N. Bohr in his rejection[7, 8] showed to be incompatible with the conceptions of the quantum theory: these hypotheses are that the world is *separable* into distinct *'elements of reality'* and also that the quantity corresponding to each element thus separated can be defined mathematically in a precise way[2].

Let us consider with EPR two observables A and B corresponding, according to the hypothesis, to the elements of reality $\mathscr{A}$ and $\mathscr{B}$. Let us assume them to be non-commutable: if the theory is complete the elements of reality $\mathscr{A}$ and $\mathscr{B}$ do not exist simultaneously since A and B do not have a precise simultaneous definition. If B is measured there is therefore destruction of the elements of reality $\mathscr{A}$, which can be explained by the use of the measuring apparatus. But this interpretation breaks down if it is possible to measure B without a measuring apparatus. In this case, A being well defined and B being measured without changing A, the elements of reality $\mathscr{A}$ and $\mathscr{B}$ must exist at the same time. There is therefore a contradiction between the claims of quantum mechanics to represent reality and the assumption that the criterion of reality adopted by EPR characterizes any complete theory.

For that, we can imagine a simple experiment. Let us take for instance the *Gedankenexperiment* set up by Bohm[2]. A zero spin molecule splits up into two atoms U and V, each having a spin $\frac{1}{2}\hbar$, the total angular momentum remaining constant. After the separation, the spins of the two atoms are evidently correlated; the wave function predicted by quantum mechanics has the form: $\psi_0 = 1/\sqrt{2}(u_+v_- - u_-v_+)$ where u and v refer to the atoms U and V respectively, and the signs $+$ and $-$ to the orientation of each spin. In the case of a proper mixing or mixing of first degree (where each atom would correspond to a well-defined state vector) we would have a combination of ψ_0 and

$$\psi_1 = \frac{1}{\sqrt{2}}(u_+v_- + u_-v_+).^1$$

Because of the correlation, the measurement of a component (x for example) of the spin of U, U_x, gives the value of V_x equal and opposite. The measurement of U_x is therefore an indirect measurement of V_x, which does not destroy the state of V. There corresponds with it an element of reality $\mathscr{A}$ which existed evidently before, at the moment of the measurement on U. But one can argue in the same way for the three components of the spin V. If the EPR postulates are correct, it follows that some elements of reality exist in the second atom V, $\mathcal{A}$, $\mathcal{B}$, $\mathcal{C}$, which correspond to a simultaneous definition of the three components of V. Now, the wave function ψ_v, according to quantum mechanics, cannot specify precisely more than one of these components since v_x, v_y and v_z do not commute. The conclusion, in the EPR perspective, is that the wave function does not provide a complete description of all the elements of the reality of atom V. It is therefore necessary to search for a complete description, in terms of hidden variables for instance, which would constitute a subjacent deterministic substratum: a quantum state would be a set of statistical states of hidden variables.

We know Bohr's refutation of the paradox[7, 8]. The implicit hypotheses in the previous reasoning, namely the separation of the elements of reality and the proper mathematical definition of the corresponding quantities, are contradictory with the fundamental hypotheses of quantum mechanics. According to Bohr, quantum mechanics concerns the interaction of 'microsystems' with the measuring apparatus and not their intrinsic characteristics. The variables are only defined accurately by reference to their interaction, especially with a measuring instrument. What the wave function ψ_0 expresses is "the propagation of correlated potentialities" (D. Bohm[2]). In other words, the mixing made up of U and V is improper (or of second degree). Keeping within a realistic perspective one must conclude that the reality for U and V is inseparable, these two systems having interacted in the past.

If we have insisted on recalling this classical example, it is because it illustrates very well the tendency of all subsequent steps, as well as the bases of the problem set. In fact, a proof that quantum theory excludes recourse to hidden variables had been put forward as early as 1932 by J. Von Neumann. But this proof did not always appear to be sufficiently convincing and was taken up again by various authors[10, 11], one of the most recent being Bell[12], who showed that all these rejections were based on too restrictive hypotheses.

The refutation of the EPR paradox, or the arguments of the type used by Von Neumann, do not remove all sense from the search for other theories hoped to be more complete than quantum mechanics, for which one is free not to accept all the basic axioms[13]. The latter being however adequate for a wide range of phenomena, the theories of hidden variables which have been put forward since [14–16] proposed the statistical reproduction of the results of quantum mechanics. (The model proposed more recently by Bohm and Bub[13] foresees some incompatibilities except in certain cases. The coupling within a certain lapse of time, which may be short, of the hidden variables with the ordinary quantum variables would produce a 'randomization' – random display – of the former which would make them unobservable after that time lapse.)

2.2. Bell's Theorem on Local Hidden Variables

The hidden variable theories generally proposed are supposed to agree with the predictions of quantum mechanics. The proof of Von Neumann, Jauch and Piron, and Gleason, only concern limited types of hidden variable theories, for which they show the incompatibility with quantum mechanics. Bell showed that the proof proposed could not be applied to *all* the possible theories of hidden variables[12]; moreover, he was able to show that if one reduced the hypotheses about these to the simple locality hypothesis, then it is possible to put in contradiction the theory of quantum mechanics and the whole family of local hidden variable theories[21]. The importance of Bell's theorem lies on the one hand in this generalization and on the other hand in that it lends itself to experimental verification. The latter thus appears equally as a test of the theory of local hidden variables and of quantum mechanics itself.

Bell's theorem bears on a thought experiment (*Gedankenexperiment*). Its extension to realizable experiments was suggested consequently[23]. Some experimental tests had been performed before to verify various predictions relating to hidden variables: they will be discussed below. They were in fact insufficient to test the predictions of the type suggested by Bell. Finer experiments were necessary. Before coming to these, we shall recall the statement of the inequalities expressed in Bell's theorem.

We take a local point of view. Two devices A and B (of the Stern and Gerlach type for instance) are separated from each other in such a way that the results of the measurement of one cannot depend on the other. Let us suppose that a be the parameter of the apparatus A and b that of B: rotational angles of polarizers for instance. Let us further suppose that the results of the measurement be of binary type (the reply is either yes or no, that is to say for example $+1$ or -1: a measure of spin orientation). We have, then:

(1) $A(a) = \pm 1, \qquad B(b) = \pm 1.$

Quantum mechanics provides the probability of having $A(a) = +1$ or -1, and the same for $B(b)$, but it cannot say anything more. Let us assume the existence of a continuous local hidden variable, λ, which completes the description, determines the phenomenon and, in particular, the correlation between A and B. Then we still have:

(2) $A(a, \lambda) = \pm 1, \qquad B(b, \lambda) = \pm 1,$

but this time the value $+1$ or -1 is determined according to the values of λ.

Let $\rho(\lambda)$ be the density of probability of λ, whose range of definition is Λ (in which case we have $\int_\Lambda \rho(\lambda)\, d\lambda = 1$). The mean values of A and B, as well as the correlation function are:

$$P_A(a) = \langle A(a, \lambda) \rangle = \int d\lambda\, \rho(\lambda) A(a, \lambda),$$

(3) $$P_B(b) = \langle B(b, \lambda) \rangle = \int d\lambda\, \rho(\lambda) B(b, \lambda),$$

$$P(a, b) = \langle A(a, \lambda) B(b, \lambda) \rangle = \int d\lambda\, \rho(\lambda) A(a, \lambda) B(b, \lambda).$$

Locality requires that $A(a, \lambda)$ be independent of b and that $B(b, \lambda)$ be independent of a.

We consider another direction for measurement corresponding to the parameter c. One can define the correlation function $P(a, c)$ according to the equation type 3. Let us introduce the determination of the correlations: for a value b' of a, the ranges Λ_+ and Λ_- of Λ are defined in such a way that:

(4)
$$\lambda \in \Lambda_+ \qquad \text{if} \quad A(b', \lambda) = +B(b, \lambda),$$
$$\lambda \in \Lambda_- \qquad \text{if} \quad A(b', \lambda) = -B(b, \lambda).$$

By simple considerations based on the relations (2) and (3), we show that:

(5) $|P(a, b) - P(a, c)| \leq 2 - P(b', b) - P(b', c).$

The inequality (5) is connected with the local hidden variables.

Quantum mechanics, for its part, predicts a correlation function

(6) $P(a, b) = \langle (\boldsymbol{\sigma} \cdot \mathbf{a})(\boldsymbol{\sigma} \cdot \mathbf{b}) \rangle = -\mathbf{a} \cdot \mathbf{b}.$

If we put (6) in (5), and if we take for example:

$\mathbf{b'} = -\mathbf{b}, \mathbf{a} \perp \mathbf{b}$, we have:

(7) $|\mathbf{a}, \mathbf{c}| < 2 - b^2 - bc.$

Let θ be the angle between the directions b and c (angle between two directions of the polarizer $\underline{B}$). (7) becomes:

(8) $|\sin \theta| \leqslant 1 - \cos \theta,$

an inequality which generally is not satisfied. For θ small, for instance, it gives $\theta \leqslant \theta^2/2$, which is wrong. There is therefore a contradiction between local hidden variable theories and quantum mechanics.

A generalization of Bell's relation has been suggested[23]. The inequality (5) which has the form:

(9) $|P(a, b) - P(a, b')| + P(a', b) + P(a', b') \leqslant 2,$

has been extended to

(10) $|P(a, b) - P(a, b')| + |P(a', b) + P(a', b')| \leqslant 2.$

(For a discussion of Bell's theorem, cf. Refs. [4, 22–26, etc....])

It is necessary to note that the above inequalities remain valid if we suppose that the hidden parameters are not the same ones for the particles occurring in A and in B [25].

2.3. *From Gedankenexperimenten to Real Experiments*

Most of the time the problem of possible tests for hidden variable theories has been dealt with in terms of gedankenexperiments. However, some experimental attempts have been made to verifications, or rather, at least up to the statement of Bell's theorem, experimental data have been interpreted as a function of these theories. That is the case for instance of Wu and Shaknov's experiment[31], studied by D. Bohm to serve as a test in a possible theory[20]. Later on, the experiments of Kocher and Commins[32], Papaliolos[33] and those which followed, aimed explicitly at the hidden variable problem. In fact, as we shall see further on, none of the mentioned experiments was precise enough with regard to Bell's predictions.

Clauser *et al.* were the first to suggest verifying Bell's relations, that is to say, to decide between quantum mechanics and local hidden variables, by applying the inequalities to a really possible experimental situation. Let us imagine that A and B each to be a photon filter[2] followed by a detector. $A(a)$ and $B(b)$ are +1 if photon detection occurs, and −1 if there is no detection.

If the counting levels are significant, then the $P(a, b)$ are given by them directly: the inequalities (5) are directly verifiable.

If the direct counting levels are too low (photo-electric efficiency very small), it is necessary to use the coincidence detection rate, $R(a, b)$, proportional to the probability that $A(a) = +1$, $B(b) = +1$, on condition that we accept the supplementary hypothesis that the simultaneous detection probability in A and B is independent of a and b. The quantities will be normalized in relation to the level of coincidence without filters in A and B. Let us take the case of an experiment with the emission of 2 γ's, each one arriving in A and B. Let $R(a, b)$ be the coincidence rate for the parameters a and b, R_0, the rate when the A and B polarizers are removed, $R_1(a)$ when the B polarizer is removed, and $R_2(b)$ when the A polarizer is removed. We then have [23]:

$$(11) \qquad P(a, b) = \frac{4R(a, b)}{R_0} - \frac{2R_1(a)}{R_0} - \frac{2R_2(b)}{R_0} + 1.$$

Introducing this into (5), we obtain the inequality between experimental quantities ($R_1(a)$ and $R_2(b)$ must normally be constant if in the absence of correlation there is no privileged direction):

$$(12) \qquad |R(a, b) - R(a, c)| + R(b', b) + R(b', c) - R_1 - R_2 \leq 0,$$

or:

$$(13) \qquad |R(\alpha) - R(\alpha + \beta)| + R(\gamma) + R(\beta + \gamma) - R_1 - R_2 \leq 0,$$

if we consider an experiment in which only the differences between the parameters a, b, etc.... are measured, by taking $\alpha \equiv b - a$, $\beta \equiv c - b$, $\gamma \equiv b - b'$, i.e.,

$$(14) \qquad |R(\alpha) - R(\alpha + \beta)| \leq 2 - R(\gamma) - R(\beta + \gamma).$$

In what follows we shall review various experiments. They will be classified according to their methods and for each method, if necessary, we shall recall the context of the tests undertaken in the light of the requirements of Bell's theorem.

3. SUCCESSIVE MEASUREMENTS OF PHOTON POLARIZATION

It is only because of the plan followed concerning the different experimental methods that Papaliolos's experiment[33] will be considered first: it has not yet been redone and its results are unsurprising. Besides, it attempts to deal with a model or rather a family of models of hidden variables, that of Bohm and Bub[13], whose problematics may seem particularly interesting (cf. also Ref. [25] pp. 177–193), and which offers itself, under certain conditions, to experimental verification, independently of Bell's inequalities.

Let us consider the model of Bohm and Bub. A system with two quantum states is made up by the photon, with its two polarization states (or a linear combination of the latter). These states can be studied by a measurement using linear polarizers. Each state being represented by $|a_1\rangle$ and $|a_2\rangle$, the wave function of the photon is $\psi = \psi_1|a_1\rangle + \psi_2|a_2\rangle$, ψ_1 and ψ_2 being complex numbers such that $|\psi_1|^2 + |\psi_2|^2 = 1$. Let us suppose that with each photon there is a similarly associated pair of hidden variables ξ_1 and ξ_2 such that $|\xi_1|^2 + |\xi_2|^2 = 1$ and that the conditions of normalization are maintained in the process which connects the ψ_i to the ξ_i during the measuring procedure. If $|\psi_1|$ and $|\xi_1|$ are known, the theory states that the result of the measurement S is completely predicted. If, just before the measurement, $|\psi_1| > |\xi_1|$, we shall have $\psi = |a_1\rangle$. If it is the opposite, i.e., $|\psi_1| < |\xi_1|$, we shall have $\psi = |a_2\rangle$. If ξ_1 is distributed uniformly in the complex plane, we obtain again the classical result of quantum mechanics.

If the measuring time is sufficiently short, the ξ_i remain more or less constant. The relaxation time of the ξ_i is situated at $\tau \sim 10^{-13}$ s, for a set of systems. This theory is therefore testable in so far as the relaxation time is sufficiently large in relation to the measuring time.

Although it contains a number of arbitrary features, this theory might be the starting point of more general considerations, such as those discussed in Ref. [39].

Experimentally one uses a set-up of three consecutive linear polarizers; the variation of transmission is measured with the rotation

M. PATY

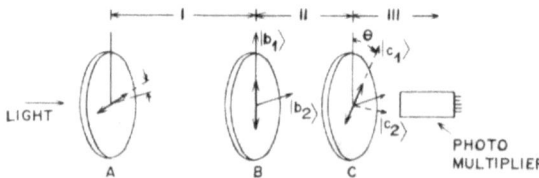

Fig. 1. *Papaliolos's experiment* (Ref. [33]). In the figure, the thick arrows indicate the
direction of polarization given by each polarizer. The arrows relative to $|b_1\rangle$ and $|b_2\rangle$
indicate the direction of polarization for the basic states of polarizer B and $|c_1\rangle$ and
$|c_2\rangle$ for the basic states of polarizer C. The angle ε is equal to $10°$. The linear polarizers
A, B, C are such that B defines the 'hidden variable' ξ_1; ξ_1 is known by the information
on the angle ε defined by A which is small:

$$|\sin \varepsilon| = |\psi| > |\xi_1|,$$

at the exit of B, ψ_1 and ξ_1 are known, therefore we know the state of the photon whose
wave function is $\psi_1 = |b_1\rangle$. The distance BC is sufficiently small for ξ_1 to be more or less
constant. We can therefore predict the ratio of the level of transmission in the region
III (see Figure 2).

of the last polarizer (that is to say with the angle θ, cf. Figure 1). The
photons, at the exit of the second polarizer B, are in a well defined
quantum state and have well defined hidden variables. C is placed
sufficiently closed to B so that ξ has not had the time to undergo its
relaxation ($t \sim 7.5 \times 10^{-14}$ s).

According to the theory, the photons which emerge from B are
such that $|\psi_1| > |\xi_1|$, ($\psi = \sin \varepsilon$). B therefore defines ξ_1. In Region II,
one knows everything about the photon: $\psi = |b_1\rangle$, $|\xi_1| < \sin \varepsilon$. (If ε is
small, one knows ξ_1 with more precision.) It can be shown that for the
photons which pass through Region III, one must have:

$$(15) \qquad \frac{\tan\left(\dfrac{\pi}{4} - \theta\right)}{1 - \tan^2\left(\dfrac{\pi}{4} - \theta\right)} > \tan \varepsilon,$$

which is the case if $\pi/4 - \theta > \varepsilon$. In other words, if $0 < \theta < \pi/4 - \varepsilon$,
transmission is certain, if $\pi/4 + \varepsilon < \theta < \pi/2$, absorption is certain.

Let $\varepsilon = 10°$
$\begin{cases} \theta < 35° \text{ there is certainly transmission in III,} \\ \theta > 55° \text{ there is certainly absorption} \\ \text{between the two, law of linearity,} \end{cases}$

θ is made to vary by steps of 10°, from 0° to 90°,
  (1) with B and C in contact,
  (2) with B and C separated by 76×10^{-4} cm.

The results (Figure 2) are in agreement with quantum mechanics to 1%, but may be compatible with the theory of Bohm and Bub if the relaxation time of the hidden variables ξ_i is short: $\tau < 2.4 \times 10^{-14}$ s (for a measuring time of 7.5×10^{-14} s).

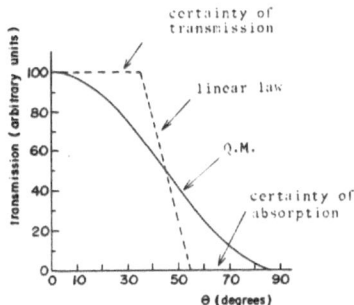

Fig. 2. *Results of Papaliolos's experiment* (taken from Ref. [33]). The curve in continuous line shows the transmission, as a function of the angle θ, predicted by quantum mechanics (proportional to $\cos^2 \theta$). The curve in broken line is that predicted by the theory of Bohm and Bub for $\varepsilon = 10°$. The data are in agreement with quantum mechanics to 1%.

However, some doubts have been expressed about the validity of the Papaliolos experiment as a test of the Bohm-Bub theory. I just quote in this respect a remark by B. Hiley[38].

The main trouble lies in the fact that he uses photons rather than particles of finite rest mass (...). In the particle case it is quite clear that the particle keeps its identity as it passes through a Stern-Gerlach magnet and it is, therefore, meaningful to assume that the hidden variables of the particle can remain unchanged as the particle passes through the Stern-Gerlach magnet. This argument does *not* hold for photons passing through polarizers. Here the process is the absorption of the photon and the re-emission in a different state of polarization at a slightly later time so that the identity of the photon is called into question. If we cannot maintain the identity of the photon in passing through a polarizer, how can we justify the assertion that the h.v's before absorption are the same as those after subsequent emission? It seems much more likely that the relation between these two sets of h.v's is, in fact, random. If this is the case, then Papaliolos' conclusions are false.

4. $\gamma\gamma$ CORRELATIONS IN THE POSITRONIUM ANNIHILATION

Let us consider the reaction $e^+e^- \rightarrow \gamma_1\gamma_2$ where the initial state has a spin 0 (s state). (It is supposed that the contribution of the state $S = 1$ is negligible.) Quantum mechanics predicts that γ_1 and γ_2 are polarized at right angles in relation to each other. The system of the 2γ is in a state $\phi = 1/\sqrt{2}(\phi_1 - \phi_2)$, with $\phi_1 = C_1^x C_2^y \phi_0$, $\phi_2 = C_1^y C_2^x \phi_0$, $\phi_0 =$ vacuum state, $C_i^k =$ creation operator of γ_i, with an impulsion directed according to z and a linear polarization depending on $k(k = x, y)$. According to quantum mechanics, ϕ_1 and ϕ_2 are orthogonal.

In a theory of hidden variables, each of the states γ_1 and γ_2 is precisely determined and not simply both together.

4.1. *Wu and Shaknov's Experiment* [31] (1950)

Cu^{64} is used as a source of positrons, produced by the excitation of Cu by deuterons (in the Columbia cyclotron). The source Cu^{64} is enclosed in lead. Two thin pipes in opposite directions channel the passage of γ to within 3°. As these γ have energies of 510 keV, their polarization can only be measured by their scattering: their Compton scattering at about 90° (in fact 82° on average) on aluminium is used. The theory predicts maximum isotropy at 82°.

Let ϕ be the azimuthal angle of scattering of the second γ in relation to the first one (Figure 3). To begin with, the first detector is held stable and ϕ is varied from 0 to 360° in steps of 90°. (Only the correlations of polarization at right angle will be tested in this way.) Then one reverses the process by fixing the second one, and varying ϕ. Series of measurements are performed alternatively in this way. The difficulty with this experiment lies in the reduction of statistical sensitivity between the linear polarization of the photon and its Compton scattering.

Let R be the ratio of the counting levels of the two counters when they are at 90° relative to each other and when they are coplanar.

For the experimental set-up adopted, quantum mechanics predicts $R_{QM} = 2$; the result of measurement is: $R_{exp} = 2.04 \pm 0.008$. Agreement with quantum mechanics is therefore excellent and in contradiction with various models of hidden variables [20].

This experiment cannot, however, tell us anything about Bell's theorem: the γ's are not subject to a binary decision. It can therefore

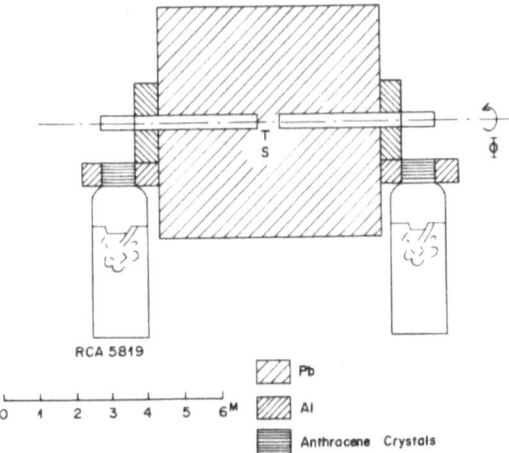

Fig. 3. *Wu's and Shaknov's experiment.* S is the source (Cu⁶⁴). The photons emitted in the channel are scattered on aluminium. They are detected at 90°. The angle ϕ varies by steps of 90°.

be considered that the same theoretical value $R_{th} = 2$ could be predicted by a theory of hidden variables (cf. e.g. Ref. [27]). Besides, it should be noted that the Compton scattering formula required by the experimental set up is itself provided by quantum mechanics (cf. Ref. [6], p. 127). In fact, most calculations made under the EPR hypothesis admit the concept of quantum mechanics which decreases the implication of the 'verification'. This will be referred to again later on.

This kind of experiment has been redone recently at Birkbeck College, the correlation being investigated over a distance of up to 2.5 m separation between the Compton scatterers. No significant deviation from quantum predictions was found [38, 40].

4.2. Kasday's Experiment [27]

The positrons emitted by a radioactive source are stopped and annihilated in copper. A given direction of γ emission is selected by using a lead collimator; the photons undergo a Compton scattering in S_1 and S_2 (see Figure 4) and arrive at the detectors D_1 and D_2, which measure their energy. The difference ϕ of the azimuthal angles ϕ_1 and ϕ_2 can be varied. Maximum count is expected, according to quantum

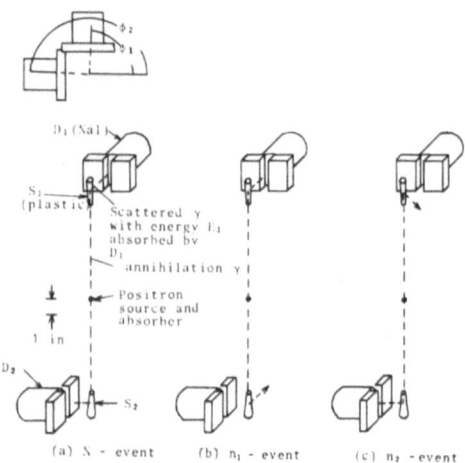

Fig. 4. *Kasday's experiment* (Ref. [27]). Diagram of the experimental set-up. The lead collimator is not shown. (a) diagram of an event at quadruple coincidence; (b) and (c) diagrams of an event at triple coincidence.

mechanics, for $\phi = 90°$ and minimal count for $\phi = 0°$; between the two it should vary as $1 - A \cos 2\phi$.

Experimentally, R is defined as a product of conditional probabilities:

$$R_{exp}(\phi_1, \phi_2) = \left(\frac{N}{N_S}\right) \Big/ \left(\frac{n_1}{N_S} \cdot \frac{n_2}{N_S}\right),$$

with N = number of times when the 2γ's undergo a Compton scattering,

NS = number of times when the 2γ's undergo a Compton scattering and a detection,

n_1 = number of times when the 2γ's undergo a Compton scattering and the first one only is detected,

n_2 = number of times when the 2γ's undergo a Compton scattering and the second one only is detected.

Figure 5 shows the results: R fits well as a function of ϕ according to the expected curve. A finer analysis may be done taking account of the distribution of the energy of the scattered photons. If attention is limited to the photons scattered without having suffered a noticeable energy loss, it is possible to distinguish between the predictions of

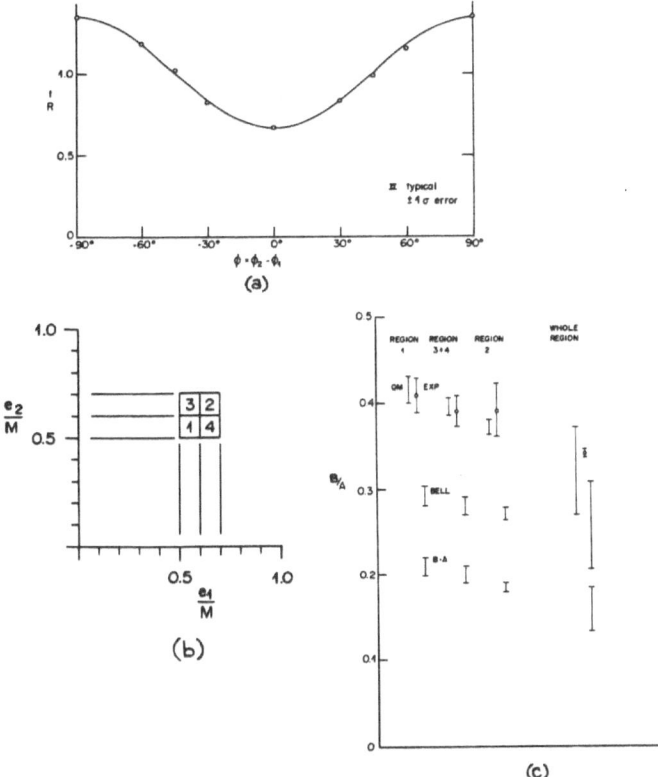

Fig. 5. *Results of Kasday's experiment* (Ref. [27]). (a) Variation of the rate R as a function of the relative azimuthal angle. The data agree with the predictions of the type $A + B \cos 2\phi$ of quantum mechanics, with A and B adjustable. (b) The energy regions chosen for the study of B/A. The quantities e_1 and e_2 are the energies of the scattered photons. M is the electron mass. (c) Comparison of the experimental results and the theoretical predictions: measurement of the ratio B/A for the regions 1, 3 + 4, 2, and the whole region. QM indicates the predictions of quantum mechanics, Bell the upper limits of Bell's inequalities, $B - A$ the upper limit of Bohm's and Aharanov's hypothesis; exp indicates the experimental results.

quantum mechanics, those of Bell's inequalities (upper limit) and those of Bohm-Aharanov. The results are then unambiguously in favour of quantum mechanics. In the absence of discrimination amongst the scattered photons, they could agree with the latter but also with Bell's inequalities (Figures 5b and c).

But the assertion of the inapplicability of any local hidden variable theory would require at the same time the certainty that the relationship between an ideal and a Compton polarizer required by quantum mechanics is true – admitting the possibility of an ideal polarizer: hence, the author cautiously concludes that his experiment cannot really settle the problem of local hidden variables.

4.3. Faraci et al.'s Experiment (Catania, 1973)[34]

Here it concerns an experiment similar to the previous one aiming to study the correlation function in terms of hidden variables. ϕ being defined as before, we have:

$$(16) \qquad W(\phi) = K \sin^2 \phi + (1 - K) \cos^2 \phi,$$

K being the correlation constant introduced[28] to express the probability of observing the pair of γ's in states of orthogonal linear polarizations. (This formula assumes ideal polarizers.) In the case of an impure mixture or of second degree (that of quantum mechanics, see above), $k = 1$; in the case of a pure mixture or of first kind (assumed to be the case for the local hidden variables), $k = \frac{3}{4}$.

More generally, if (16) is compatible with Bell's inequalities (9, 10), one finds that

$$(17) \qquad \frac{1}{7} \approx \frac{1}{\sqrt{2}} - \frac{\sqrt{2}}{4} < K < \frac{1}{2} + \frac{\sqrt{2}}{4} \approx \frac{6}{7}.$$

The polarization measurement of the γ's is in fact obtained by Compton scattering: the experimental set up is shown in Figure 6. Using the empirical formula of Klein-Nishijima for the Compton scattering cross-section, we obtain the correlation function of γ_1 and γ_2 in coincidence, which replaces formula (16):

$$(18) \qquad W(\theta_1, \theta_2, \phi) \sim K_1^2 K_2^2 \{\gamma_1\gamma_2 - \gamma_1 \sin^2 \theta_2 - \gamma_2 \sin^2 \theta_1 \\ + 2 \sin^2 \theta_1 \sin^2 \theta_2 [(2K - 1) \sin^2 \phi + (1 - K)]\},$$

with $\gamma_i = K_i/K_0 + K_0/K_i$; L_0 and K_i: the wave numbers of entering and leaving γ's ($i = 1, 2$); θ_1 and θ_2: scattering angles of γ_1 and γ_2; ϕ: the azimuthal angle.

We measure the coincidence counting rate $N(\theta_1, \theta_2, \phi)$ between the scintillators S_1, S_2 (plastic scintillators, Compton scatterers) and R_1 and R_2 (NaI scintillators). The source of e^+ is ^{22}Na surrounded by

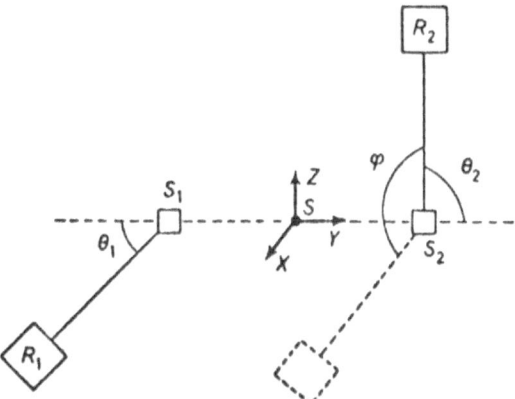

Fig. 6. *Faraci et al.'s experiment* (Ref. [35]). Diagram of the experimental set-up. S_1 and S_2 are Compton scintillators and scatterer. R_1 and R_2 (crystals of NaI) are placed at angles θ_1 and θ_2. Na^{22} represents the source S.

plexiglass which annihilates the e^+. The resolution time in coincidence is 30 ns. $N(\theta_1, \theta_2, \phi)$ constitutes a measure of $W(\theta_1, \theta_2, \phi)$, allowing for a correction for the effects of finite geometry. Figure 7 compares the experimental results with the theoretical predictions for $N(\theta, \theta, 90°)/N(\theta, \theta, 0°)$; in Figure 8 the ratio $N(60°, 60°, \phi)/N(60°, 60°, 0°)$ is shown. The conclusion is that the results disagree with the predictions of quantum mechanics, and are in agreement with the upper limit of Bell's inequalities.

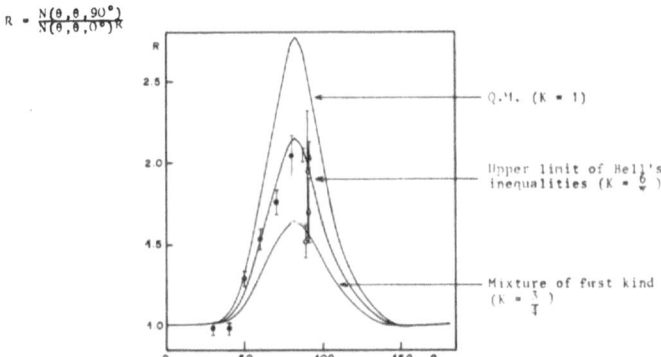

Fig. 7. *Faraci et al.'s experiment* (Ref. [35]). Anisotropy ratio as a function of the scattering angle. The theoretical curves are indicated in the figure.

$$R_{(\phi)} = \frac{N(60°, 60°, \phi)}{N(60°, 60°, 0°)}$$

Fig. 8. *Faraci et al.'s experiment* (Ref. [35]). Correlation of the direction as a function of the azimuthal angle.

It should be noted that these results are half way between predictions related to mixture of first and second kind; since they are found just on the (upper) limit of Bell's inequalities, they cannot give a precise answer concerning the existence or non-existence of hidden variables.

Figure 9 represents the variations of the anisotropy ratio $R = N(60°, 60°, 90°)/N(60°, 60°, 0°)$ for distances $\lambda_i = S - S_i$ being different. It is possible, indeed, to suppose that there exists an effect as a function of the distances. For example [20], if the two photons, still unseparated constitute a mixture of second kind, one can admit that a (hidden) process means that a well defined state of polarization corresponds to each one of them, as soon as they are sufficiently separated (the states of polarization of course being related to one another, given their past relationship). Since the distribution of the whole is supposed uniform along all the individual directions, we obtain in fact a statistical result. This type of explanation could

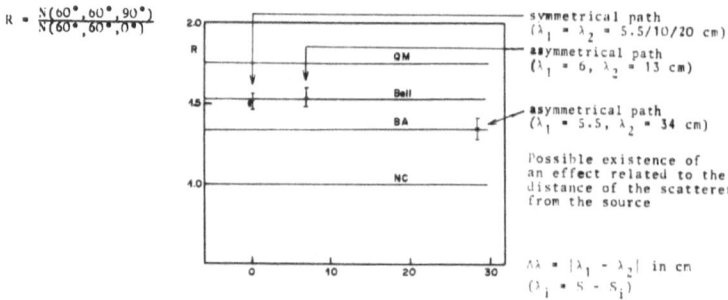

Fig. 9. *Faraci et al.'s experiment* (Ref. [35]). Anisotropy ratio $\theta_1 = \theta_2 = 60°$ as a function of the path difference of the two photons.

account for the effect quoted in Figure 9. For very asymmetrical λ_i ($\lambda_i = 5.5$ cm, $\lambda_2 = 34$ cm), we are in agreement with a mixture of first kind. It is evident that these kinds of effects should be tested in great detail in the future.

We shall further discuss this experiment in our conclusion.

An independent experimental investigation has been performed recently at Birkbeck College, which does not find the quoted change in the asymmetric case of Figure 9, and points out that important geometrical and instrumental considerations might affect the results of Faraci et al.[40, 41].

5. $\gamma\gamma$ CORRELATIONS IN THE ATOMIC DE-EXCITATION IN CASCADE

5.1. Kocher's and Commins's Experiment [32]

Figure 10 shows the de-excitation cascade of calcium

$$J = 0 \xrightarrow{\gamma_1} J = 1 \xrightarrow{\gamma_2} J = 0,$$

the γ_1 and γ_2 photons are emitted in the visible spectrum. The polarization correlation can be measured with linear polarizers such as polaroid and photomultipliers of current use.

Figure 11 shows the principle of the experimental set-up (it is in fact relative to the more sophisticated further experiment of Freedman and Clauser). A calcium beam produced by a tantalum oven is excited by an H_2 arc lamp (a D_2 lamp in the later experiments[34]). About 10% of the atoms excited at the 6^1P level go

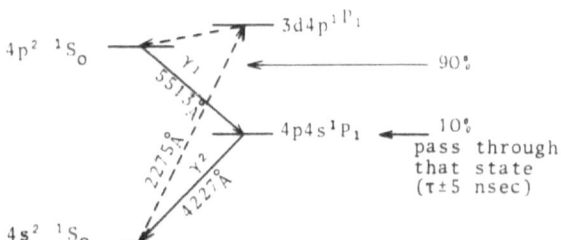

Fig. 10. *Freedman and Clauser's experiment* (Ref. [34]). Diagram of the calcium levels. The dotted line indicates the excitation process up to $4p^{21}S_0$.

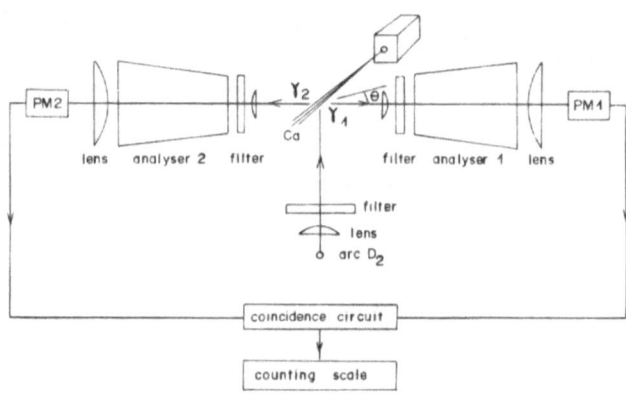

Fig. 11. *Ibid.*; diagram of the experimental set-up.

to the 6^1S_0 state (the other 90% returning to ground state directly). Without polarizers the coincidence rate is $\sim 10^{-1}\,s^{-1}$.

Measurements are made with the axes of the polarizers alternatively parallel and perpendicular. The correlation depends only on the relative angles between the axes of polarization. Figure 12 shows the results obtained. Because of the character of the

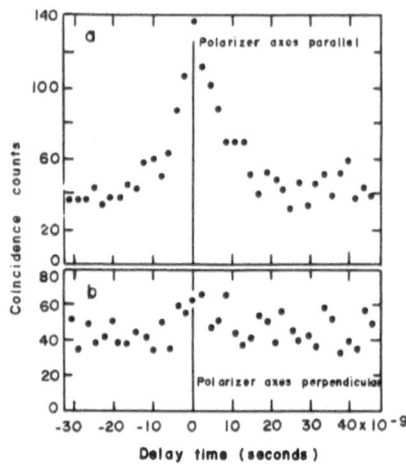

Fig. 12. *Kocher and Commins's experiment* (Ref. [32]). Coincidence counting rate as a function of time, showing the correlation of polarization.

transitions produced by and resulting in states of same spin (0) and parity, the counting rate predicted by quantum mechanics is proportional to $(\varepsilon_1 \cdot \varepsilon_2)^2$, that is to say that the polarizers are parallel. This is exactly what is observed in Figure 12.

However, the polarizers have only a relative efficiency and the two orientations of 0° and 90° do not allow a finer analysis using, for example, Bell's inequalities.

5.2. Experimental Conditions for a Conclusive Experiment

This experiment is based on the same principle as the previous one, but allows one to test Relations (12) and (13). Clauser et al.[23] have calculated the efficiencies that polarizers 1 and 2 should possess to make it possible to test these relationships. For an electric dipolar cascade of the type described before, the counting rate predicted by quantum mechanics, $R(\phi)/R_0$ (ϕ being the angle between the axes of polarizers), is given as a function of θ (half opening angle of the detectors), of ε_M^i (efficiency of polarizer i, $i = 1, 2$, when the light is polarized in parallel with the axis of polarization) and ε_m^i (efficiency of the polarizer i, when the light is polarized perpendicularly to the axis):

$$\frac{R(\phi)}{R_0} = \tfrac{1}{4}(\varepsilon_M^2 + \varepsilon_m^1)(\varepsilon_M^2 + \varepsilon_m^2)$$

(19)
$$+ \tfrac{1}{4}(\varepsilon_M^1 - \varepsilon_m^1)(\varepsilon_M^2 - \varepsilon_m^2)F_1(\theta)\cos 2\phi,$$

$$\frac{R_i}{R_0} = \tfrac{1}{2}(\varepsilon_M^i + \varepsilon_m^i) \qquad (i = 1, 2).$$

$F_1(\theta)$ is a function which represents the depolarization due to the non-collinearity of the photons. It is equal to 1 for an infinitesimal opening angle.

In order to show that (13) is violated, it is merely necessary to have efficient polarizers. The greatest degree of violation is obtained for $\alpha = 22°5$, $\beta = 45°$ and $\gamma = 157°5$ in the case under consideration.

The definition of the violation of (13) is:

$$\sqrt{2}F_1(\theta) + 1 > \frac{2}{\varepsilon_M}.$$

For a given $F_1(\theta)$ it is therefore necessary to have a minimum value of ε_M. A lower limit of $F_1(\theta)$ corresponds to an upper limit of θ: the

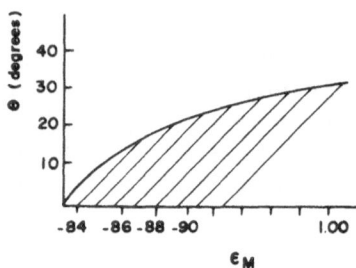

Fig. 13. Definition of the experimental requirements for testing Bell's inequalities (according to Ref. [23]). See text, Section 5.2.

region below the curve (Figure 13) characterizes a conclusive experiment.

5.3. *Freedman and Clauser's Experiment*

The experimental conditions described above were realized in Freedman's and Clauser's experiment[34]. Figure 11 shows the experimental set-up. The polarizers are 'piled plates'; 10 glass sheets, 0.3 mm thick, are inclined at Brewster incidence. They can be rotated and turned in steps of 22°5. Their efficiencies are:

$$\varepsilon_M^1 = 0.97 \pm 0.01, \qquad \varepsilon_m^1 = 0.038 \pm 0.004,$$

$$\varepsilon_M^2 = 0.96 \pm 0.01, \qquad \varepsilon_m^2 = 0.037 \pm 0.004.$$

The opening angle is 30°; the corresponding function takes the value $F_1(30°) \sim 0.99$.

The measurements with rotations and the measurements without polarizers are alternated (the polarizers being removable).

Photomultipliers are situated behind the polarizers to detect γ_1 and γ_2; an electronic circuit assures the coincidence (the resolution time is of 1.5 nsec; taking account of the fact that the intermediate state of calcium has a mean life of $\tau = 5$ nsec, the coincidence window can be as small as 8 nsec).

The experiment requires a long run, due to the low counting rate (0.3–0.1 count per second without polarizers): the results refer to 200 h of run.

Let us consider some local hidden variable theory. It can be reduced to the hypothesis that the two photons are propagated as two localized and separated particles; that each of them undergoes at each polarizer a binary selection (with transmission or not) and that this does not depend on the other polarizer. Besides, let us suppose that the probability of detection is independent of whether or not the photon passes through a polarizer. These hypotheses are equivalent to supposing the inequalities (characteristic of local hidden variable theories):

$$(20) \qquad -1 \leqslant \Delta(\phi) \leqslant 0 \qquad \text{with} \quad \Delta(\phi) = \frac{3R(\phi)}{R_0} - \frac{R(3\phi)}{R_0} - \frac{R_1 + R_2}{R_0}.$$

With θ sufficiently small and high efficiency of the polarizers, (19) and (20) are contradictory for a certain region of ϕ. The maximal violation of (20) is expected for

$$\phi = 22°5 \qquad (\Delta(\phi) > 0) \quad \text{and} \quad \phi = 67°5 \qquad (\Delta(\phi) < -1).$$

For these values, (20) is equivalent to (21):

$$(21) \qquad \delta = \left| \frac{R(22°5)}{R_0} - \frac{R(67°5)}{R_0} \right| - \frac{1}{4} \leqslant 0.$$

Figure 14 shows the result obtained. The experimental points are in agreement with the predictions of quantum mechanics. The quantity δ is: $\delta_{\text{exp}} = 0.050 \pm 0.008$, in contradiction with (21).

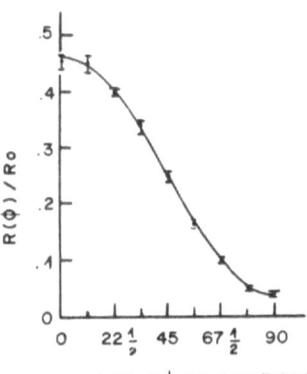

ANGLE ϕ IN DEGREES

Fig. 14. *Freedman and Clauser's experiment* (Ref. [34]). Rate of coincidence (normalized) as a function of the angle between the polarizers. The curve shows the predictions of quantum mechanics corrected for the experimental conditions (efficiency of a polarizer and limitations of solid angle).

This experiment, therefore, leads us to the conclusion that local hidden variables are absent.

5.4. *Holt's Experiment*

This time the de-excitation in a mercury atom ^{198}Hg is used. The level diagram is represented in Figure 15. It should be noted that here it

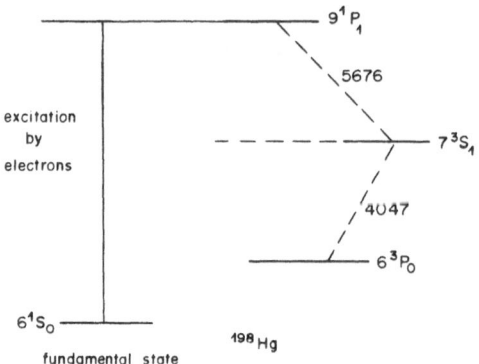

Fig. 15. *Holt's experiment* (according to Ref. [36]). 1-1-0 Mercury transition: $9^1P_1 \rightarrow 7^3S_1 \rightarrow 6^3P_0$. 7^3S_1 transition state is at $J = 0$; $M = 1, 0, -1$. (3 degenerate states). The transition passes only through the states $M = \pm 1$ (since $\Delta M = 1$).

concerns the first observation of this type of transition, and furthermore the same experiment has made it possible to determine the lifetime of the transition level 7^3S_1 (estimated at 8.2 ± 0.2 nsec). The mercury atoms are excited to the level $^1 9P_1$ using electrons of about 100 eV (collision time about $\leqslant 10^{-14}$ s). The experimental set-up is sketched in Figure 16.

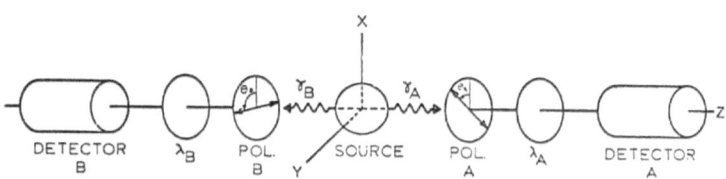

Fig. 16. *Diagram of Holt's experimental set-up* (taken from Ref. [36]).

Bell's inequalities are defined here as:

$$-\frac{1}{4} \leqslant \frac{R(67°1/2) - R(22°1/2)}{R_0} \leqslant \frac{1}{4} = 0.250.$$

Quantum mechanics predicts:

$$\frac{R_{QM}(67°1/2) - R_{QM}(22°1/2)}{R_{0QM}} = 0.269.$$

The experimental result (with a 150 h run) is as follows:

$$\frac{R(67°1/2) - R(22°1/2)}{R_0} = 0.216 \pm 0.013.$$

The correlation observed is therefore smaller than that predicted by quantum mechanics: the experimental results differ from the latter by four deviations and are found within the limits permitted by Bell's inequalities.

An extremely sophisticated study of the possibility of systematic errors has been made: it would be too long to discuss them here in detail. The author states in conclusion that "the statistical precision is certainly enough to conclude that there is disagreement with quantum mechanics". More recent results throw, however, some doubt on Holt's conclusion[42, 43] (see note added in proof on p. 288).

6. QUANTUM MECHANICS OR HIDDEN VARIABLES

To sum up, the experiments concerned with the verification of quantum mechanics, with reference to Bell's inequalities for the local hidden variable theories, provide contradictory results. Amongst the experiments studied in this outline, two favour quantum mechanics, whereas two others seem to contradict it. This contradictory situation is found again in the two principal types of experiments considered: those dealing with the correlations of polarization of photons emitted in the annihilation e^+e^- (for quantum mechanics: Kasday; against: Faraci et al.); those on the correlations of photons emitted in atomic de-excitations in cascade (for quantum mechanics: Freedman and Clauser; against: Holt).

An experiment of another type concerned with the spin correlation in proton-proton scattering, of which the results are presented in this colloquium, shows itself in agreement with quantum mechanics[37].

The present balance sheet of the experiments designed to test Bell's inequalities is therefore as follows: three agree with quantum mechanics, and two disagree. The situation is therefore new with respect to the preceding unanimity concerning the predictions of quantum mechanics in its most varied processes. Has quantum mechanics now revealed its limitations, or more exactly, the limits of its field of application? This would not be unthinkable a priori, if it is true that every theory relates to a limited field – wide as these limits might be. This would also be the case for a theory as powerful as quantum mechanics, which itself is highly powerful, but at the same time probably has a frail basis.

However, it may seem doubtful that such an established theory might be questioned in such simple experiments. And in fact quantum mechanics may only *appear* to be frail; its hold on our conceptions is paradoxically shown in this recent questioning: it is not quantum mechanics which is put into doubt, so much as the basis of these very experiments or at least their interpretation.

Amongst the possible criticisms of the two experiments which contradict the predictions of quantum mechanics, those concerning Faraci *et al.*'s experiment are the strongest. To sum them up, let us simply note that the initial state of the system e^+e^- is supposed to be such that $s = 0$, which, although reasonable, has not actually been observed, in the present case at least; the photon polarization has not been directly measured by means of an experiment of a binary type, but by Compton scattering, which introduces a transfer of information which may modify entirely the properties of the initial state (i.e. quantum mechanics is introduced by the Klein-Nishijima formula). Finally, last but not least, this experiment may present systematic errors which have not necessarily been studied.

As regards the other experiment, that of Holt, it may, at first, seem surprising that it should contradict Freedman's and Clauser's similar experiment. How can such contradictory results be explained? Is it because of the difficulties inherent to this type of experiment? The systematic experimental errors have been studied very carefully and all danger on this side seems to have been removed. But, here again, the initial state, as well as the cascade process, may not be as simple and straightforward as one might expect. The atomic cascade of mercury, described by Holt, has been studied for the first time and of course this result should be corroborated by other experiments. Besides, the de-excitation transition level of the mercury atom is degenerate and it is not impossible that some effects might result from this.

As a matter of fact, it is not impossible either that this process may actually bring to light the domain of validity of quantum mechanics.

The most immediate conclusion is obviously that one should wait for further experimental results, and probably more refined experiments.

It is anyway interesting to note that 50 years after the foundation of quantum mechanics, its most basic foundations can still be questioned, and that local hidden variable theories which reintroduce determinism in its classical sense, are not pushed away for good. However, the hidden variable theories put forward, which might lend themselves to deterministic predictions as well as to their theoretical and experimental refutation, are all founded, as it has been pointed out[29], upon an extremely vague notion of hidden variables; they are in a way nothing but 'ghosts' no doubt capable of bringing determinism back, but without any real content. In a way they are experiments in pure reasoning. The experiments described are typical experiments of quantum mechanics; in particular, one ignores in each of these experiments the initial conditions of the system and much can occur which is in fact random. Beside, these experiments do not represent really direct tests, the fundamental process being always mediated by other phenomena (belonging for instance to the equations of propagation, which introduce extra elements and 'distort' the initial phenomenon) so that one can wonder to what extent this type of experiment is really convincing.

Therefore, it seems to me that one is brought back after all, despite the definite interest in these experiments of verification which, once improved in their principle, may one day become crucial experiments, to the 'epistemological' discussion, as old as quantum mechanics, on its interpretation, the significance of its concepts, and the possible insufficiency of those we have at the moment which may still be too closely linked to classical physics.

I would like to thank Professors B. d'Espagnat, C. Imbert, G. Faraci, B. Hiley and R. Lestienne for information and fruitful discussions in relation to this work. I acknowledge the help of F. Hopkins in translating this report into English.

Centre de Recherches Nucléaires,
Université Louis Pasteur,
Strasbourg

Note added in proof: Since the completion of this work, new experiments have been performed on the same grounds as those described in the text: one with e^+e^- annihilation[40] (cf. also [41]), two with photons from cascade in mercury atoms[42, 43]. All three are in agreement with QM and violate Bell's inequalities. Thus, the score is presently six against two in favour of QM. This modification leaves our conclusions unchanged.

NOTES

[1] ψ_1 is the component of the wave function for $J = 1$ and $J_z = 0$. An improper mixing, or of second degree, is defined, on the contrary, as a mixing of systems not having well defined state vectors (which will, as we shall see, be the case here).
[2] 'Filter' refers here to the polarization state: in this particular case it is indeed a polarizer.

BIBLIOGRAPHY

[1] Einstein, A., Podolsky, B., and Rosen, N., *Phys. Rev.* **47** (1935), 777.
[2] Bohm, D., *Quantum Theory*, Prentice-Hall, 1951, pp. 611–623.
[3] d'Espagnat, B., *Conceptions de la Physique Contemporaine*, Hermann, 1965, pp. 42–50.
[4] d'Espagnat, B., *Conceptual Foundations of Quantum Mechanics*, Benjamin, 1971, pp. 99–138.
[5] Jauch, J. M., *Foundations of Quantum Mechanics*, Addison-Wesley, 1968, pp. 185–191.
[6] Hooker, C. A., in R. G. Colodny (ed.), *Paradigms and Paradoxes, The Philosophical Challenge of the Quantum Domain*, Univ. Pittsburgh Press, 1972, pp. 69–152.
[7] Bohr, N., *Phys. Rev.* **48** (1935), 696.
[8] Furry, W. H., *Phys. Rev.* **49** (1936), 393.
[9] von Neumann, J., *Mathematische Grundlagen der Quanten-Mechanik*, Berlin, 1932 (Engl. transl. Princeton, 1955). Cf. Albertson, J., *Ann. J. Phys.* **29** (1961), 478.
[10] Jauch, J. M. and Piron, C., *Helv. Phys. Acta* **36** (1963), 827.
[11] Gleason, A. M., *J. Math. and Mech.* **6** (1957), 885.
[12] Bell, J. S., *Rev. Mod. Phys.* **38** (1966) 447.

[13] Bohm, D. and Bub, J., *Rev. Mod. Phys.* **38** (1966), 453.
[14] Bohm, D., *Phys. Rev.* **85** (1952), 166; Bohm, D. and Vigier, J. P., *Phys. Rev.* **96** (1954), 208.
[15] de Broglie, L., *Compt. Rend.* **183** (1926), 447; **184** (1927), 273; **185** (1927), 380; de Broglie, L., *Étude critique des bases de l'interprétation actuelle de la mécanique ondulatoire*, Gauthier-Villars, Paris, 1963, (Engl. transl. Elsevier, Amsterdam, 1964); etc.
[16] Wiener, N. and Siegel, A., *Nuovo Cim. Suppl.* **2** (1955).
[17] Fürth, R., *Zeits. Phys.* **81** (1933) 143.
[18] Fényes, I., *Zeits. Phys.* **132** (1952), 81.
[19] Weizel, W., *Zeits. Phys.* **134** (1953), 264; **135** (1953), 270; **136** (1954), 582.
[20] Bohm, D. and Aharanov, Y., *Phys. Rev.* **108** (1957), 1070; *Nuovo Cim.* **17** (1960), 964.
[21] Bell, J. S., *Physics 1* (1965), 195.
[22] de Broglie, L., *Compt. Rend.* **278** (1974), B7, 21.
[23] Clauser, J. F., Horne, M. A., Shimony, A., and Holt, R. A., *Phys. Rev. Let.* **23** (1969), 880.
[24] Selleri, F., *Lett. Nuovo Cimento* **3** (1972), 581.
[25] Capasso, V., Fortunato, D., and Selleri, F., *Riv. Nuovo Cimento* **2** (1970), 149.
[26] Wigner, E. P., cited in Ref. [25].
[27] Kasday, L., in *Foundation of Quantum Mechanics*, Proc. 'Enrico Fermi' Summer School IL, Academic Press, New York, 1971.
[28] Jauch, J. M., *ibid.*
[29] Kershaw, D. S., prepr. Univ. Maryland, 74-034, 1974.
[30] Bell, J. S., in *Foundations of Quantum Mechanics*, *op. cit.*
[31] Wu, C. S. and Shaknov, I., *Phys. Rev.* **77** (1950), 136.
[32] Kocher, C. A. and Commins, E. D., *Phys. Rev. Lett.* **18** (1967), 575.
[33] Papaliolos, C., *Phys. Rev. Lett.* **18** (1967), 622.
[34] Freedman, S. J. and Clauser, J. F., *Phys. Rev. Lett.* **28** (1972), 938.
[35] Faraci, G., Gutkowski, D., Notarrigo, S., and Pennisi, A. R., *Let. Nuovo Cim.* **9** (1974), 607.
[36] Holt, R. A., Thesis, Harvard, 1973.
[37] Lamehi-Rachti, M. and Mittig, W., communication to this Colloquium.
[38] Hiley, B., private communication (letter to the author, 19.8.75).
[39] Baracca, A., Bohm, D., Hiley, B. J., and Stuart, A. E. G., *Nuovo Cim.* **28B** (1975), 453.
[40] Wilson, A. R., Lowe, J., and Butt, D. K., *J. Phys. G* **2** (1976), 613.
[41] Bohm, D. and Hiley, B. J., *Nuovo Cim.* **35B** (1976), 137.
[42] Clauser, J. F., *Phys. Rev. Lett.* **36** (1976), 1223.
[43] Fry, E. S. and Thompson, R. C., *Phys. Rev. Lett.* **37** (1976), 465.

M. LAMEHI-RACHTI AND W. MITTIG

SPIN CORRELATION MEASUREMENT IN PROTON-PROTON SCATTERING AND COMPARISON WITH THE THEORIES OF THE LOCAL HIDDEN VARIABLES

ABSTRACT. Bell has shown ('Bell's inequality') that local hidden variable theories lead to predictions in contradiction with quantum mechanics. This has been tested in low energy proton-proton scattering by the simultaneous measurement of the polarization of the two protons. The results are in agreement with quantum mechanics and thus in contradiction with the inequality of Bell.

1. INTRODUCTION

Some experiments[1, 5] consisting of the measurement of spin correlations of photons have been performed in order to test the validity of the local theories of hidden variables. Following the suggestion of Fox[6], we have measured the spin correlations in an experiment of proton-proton scattering.

Contrary to the preceding ones, this experiment was realized with non-zero rest mass particles, therefore localizable by a Lorentz transformation. Another difference is the participation of nuclear forces in the preparation of the state and in the polarization measurements.

2. THE PREDICTIONS OF QUANTUM MECHANICS AND LOCAL HIDDEN VARIABLE THEORIES

Let us consider the generalized inequality of Bell[7]

(1) $|P(\mathbf{a}, \mathbf{b}) - P(\mathbf{a}, \mathbf{c})| + |P(\mathbf{b}, \mathbf{b}') + P(\mathbf{b}', \mathbf{c})| \leq 2,$

where $P(\mathbf{a}, \mathbf{b})$ is a correlation function, the spin of one of the particles being measured along the axis $\mathbf{a}$ and that of the other in the direction of the axis $\mathbf{b}$. Quantum mechanics anticipates, if the two protons are

J. Leite Lopes and M. Paty (eds.), Quantum Mechanics, a Half Century Later, 291–303. All Rights Reserved
Copyright © 1977 by D. Reidel Publishing Company, Dordrecht-Holland

in a singlet state, that:

$$P_{QM}(\mathbf{a}, \mathbf{b}) = \langle \sigma_1 \mathbf{a} \sigma_2 \mathbf{b} \rangle = -\cos(\mathbf{a}, \mathbf{b}).$$

If the vectors $\mathbf{a}$, $\mathbf{b}$, $\mathbf{b}'$, $\mathbf{c}$ are in the same plane, only the angles between these orientations come in and according to (1), we shall have:

$$|P(\theta) - P(\theta + \gamma)| + |P(\phi) + P(\phi + \gamma)| \leq 2.$$

In the cases $\theta = 30°$, $\gamma = 120°$ and $\phi = 90°$, $\theta = 45°$, $\gamma = 90°$ and $\phi = -45°$ and $\theta = 60°$, $\gamma = 60°$ and $\phi = 0°$ we shall have:

$$|P(30°)| \leq \tfrac{2}{3}, \quad |P(45°)| \leq \tfrac{1}{2} \quad \text{and} \quad 3|P(60°)| \leq 2 - |P(0)|,$$

respectively.

Supposing the above correlation is perfect at $0°$ that is if $P(0°) = -1$, we shall have (Table I):

TABLE I

$|P(\phi)|$ ϕ	Predictions of QM	Bell's upper limit
0°	1	1
30°	$\sqrt{\tfrac{3}{2}}$	$\tfrac{2}{3}$
45°	$\sqrt{\tfrac{2}{2}}$	$\tfrac{1}{2}$
60°	$\tfrac{1}{2}$	$\tfrac{1}{3}$
90°	0	0

3. THE PRINCIPLE OF THE EXPERIMENT

Fox had proposed a proton-proton scattering experiment at low energy (from 1 to 4 MeV) and the measurement of the polarizations with a helium polarimeter. Due to the problems of mechanical strength in the construction of a helium polarimeter and of a gaseous target geometry, we have constructed a carbon polarimeter. For protons of 4.5 to 6.250 MeV and at a scattering angle of 50°, carbon[8] has a high analysing power which makes its use possible in spin measurements.

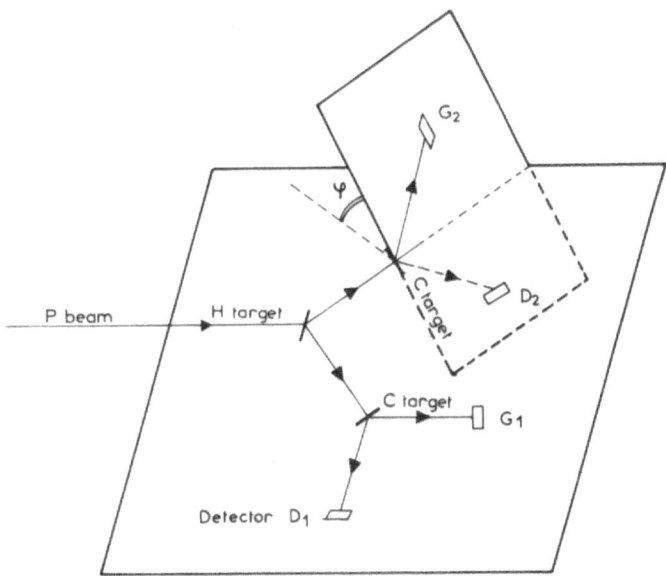

Fig. 1. Diagram of the set-up of the experiment.

Figure 1 shows the schematic arrangement of the apparatus. The experiment consists of the measurement of the correlation function:

$$F(\phi) = \frac{-D_1D_2 - G_1G_2 + D_1G_2 + G_1D_2}{D_1D_2 + G_1G_2 + D_1G_2 + D_2G_1},$$

with one polarimeter in the scattering plane and the other polarimeter turned at an angle ϕ with respect to scattering plane. D_1D_2 represent the coincidence rates between the detectors D_1 and D_2 etc.

The scattering angle of proton on proton is 45° in the laboratory (90° in the centre of mass). At the energy of the experiment (between 13 and 14 MeV in the laboratory) the quasi-totality of the scattering is in a singlet state. Singlet and triplet scattering being incoherent, at $\theta_{lab.} = 45°$, the coefficient C_{nn} can be written:

$$C_{nn} = \frac{-(d\sigma/d\Omega)_S + (d\sigma/d\Omega)_t}{(d\sigma/d\Omega)_S + (d\sigma/d\Omega)_t}.$$

C_{nn} has been measured at energies close to those of our experiment with a similar arrangement[9] and by the scattering of polarized protons on a polarized target [10]. The contribution of the triplet state can be estimated at 2 or 3%.

4. THE EXPERIMENTAL DEVICE

The efficiency of the polarimeters being approximately 10^{-5} and the detected protons having a low energy, it is necessary to obtain good energy spectra of protons with a complete separation between the interesting protons and the background. It was possible to considerably reduce the background due to gamma-rays and to neutrons by lead protection, diaphragms and an appropriate thickness of silicon detectors.

Figure 2 shows an example of proton energy spectrum detected (direct spectrum). A sheet of polyethylene $(CH_2)_n$ is used as the

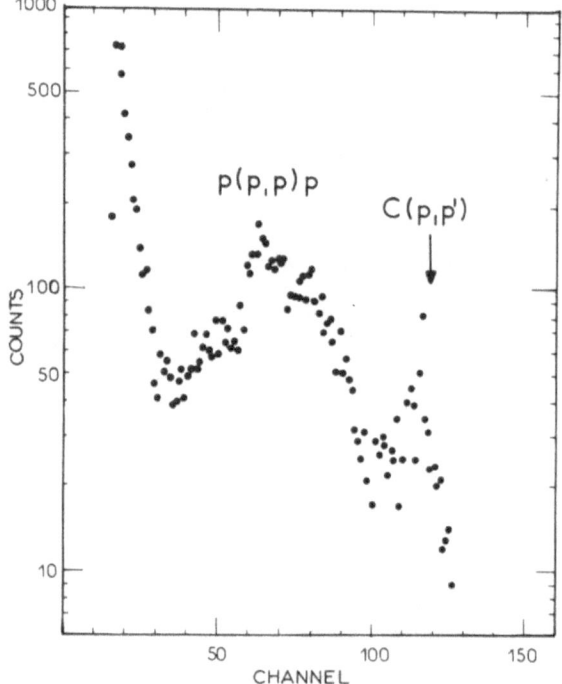

Fig. 2. Single spectrum of the protons arriving on one of the detectors.

hydrogen target. It was necessary to construct a rotating target to avoid the immediate destruction of the target by the intense beam ($\geq 1\ \mu$A) necessary in the experiment.

Figure 3 shows the arrangement of the experiment, with the hydrogen target removed.

Fig. 3. The set-up of the experiment, with the hydrogen target removed: (1) diaphragms defining the beam; (2) the lead collimator; (3) polarimeter in the reaction plane; and (4) four detectors polarimeter allowing a simultaneous measurement for ϕ and $\pi/2 - \phi$. The lead collimator and the carbon target have been removed.

The electronic equipment used, schematized in Figure 4, allows permanent recording during the experiment of the energy information about the particles falling on the two polarimeters, the time information between pulses and identification of detectors.

Figure 5 gives an example of the time spectrum obtained; its full width is 1.2 nsec at half-height. Throughout the experiment the width of the time spectrum varied very little.

Figure 6 shows an energy spectrum of protons in coincidence; it is noticeable that the γ-ray background and the inelastic peak of proton on carbon have completely disappeared.

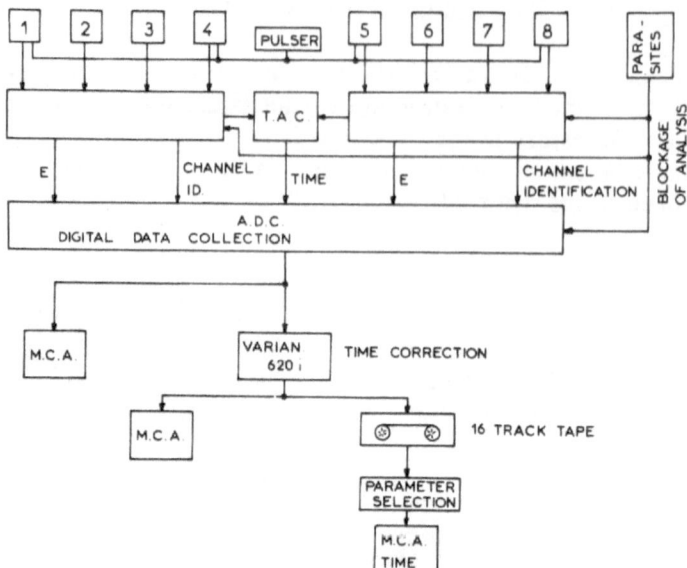

Fig. 4. Diagram of the electronic set-up.

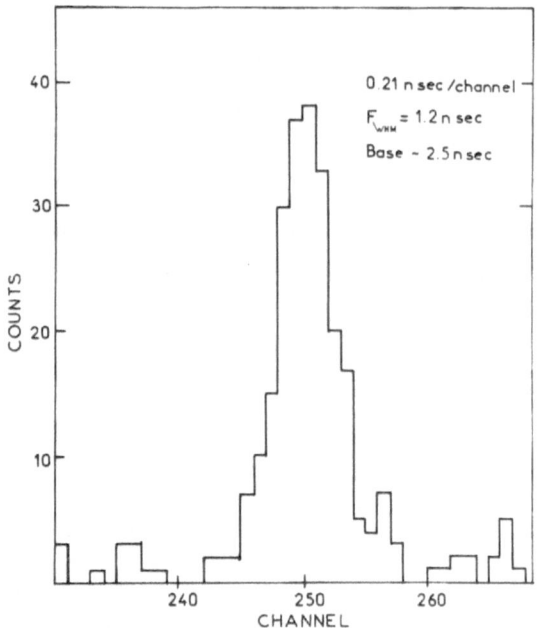

Fig. 5. Example of a time spectrum.

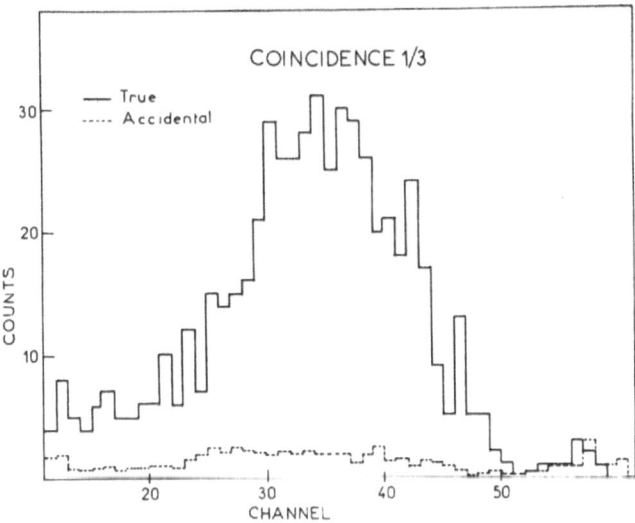

Fig. 6. A proton spectrum in coincidence.

5. MEASUREMENTS OF ANALYSING POWER AND GEOMETRICAL CORRELATION

The correlation function can be written in good approximation:

$$(2) \qquad F_{mes}(\phi) = NF_{exp}(\phi) + C_g(1 - N^2 F^2_{exp}(\phi)) \cos \phi,$$

where $N = \overline{P_A P_B}$ is the mean value of the product of the analysing power of the polarimeters, C_g is a geometrical correlation due to the anisotropy of the $^{12}C(p, p)$ scattering cross-section and $F_{exp}(\phi)$ is the experimental correlation function which equals $C_{nn} \cos \phi$ for QM or must satisfy the Bell's inequalities.

To be able to compare the experimental results with the theoretical predictions, it is necessary to take into account the geometry of the experiment (the solid angles, the diameter of the beam on the target, the thickness of the hydrogen and carbon targets etc.).

A calculation based on the Monte-Carlo method is in progress; its estimates will be able to be compared with the results of the experiment and will permit the study of the influence of the various geometrical factors.

The measurement of the analysing power and the estimate of the geometrical correlation allows the deduction from Formula (2) of the experimental correlation function, which governs the experiment.

5.1. *The Analysing Power*

The analysing power was measured in a double diffusion arrangement. A beam of protons of 8.07 MeV was scattered by a 2 mg cm^{-2} carbon target. The scattered proton enters the polarimeter at $\theta_{lab.} = 70°$. At this energy and this angle, the analysing power of the carbon is 100% [8, 11]. Polyethylene foils are put between the first and second target to slow down the protons to the desired energy. The results of the asymmetry measurements and the comparison with the Monte-Carlo calculation are shown on Figure 7.

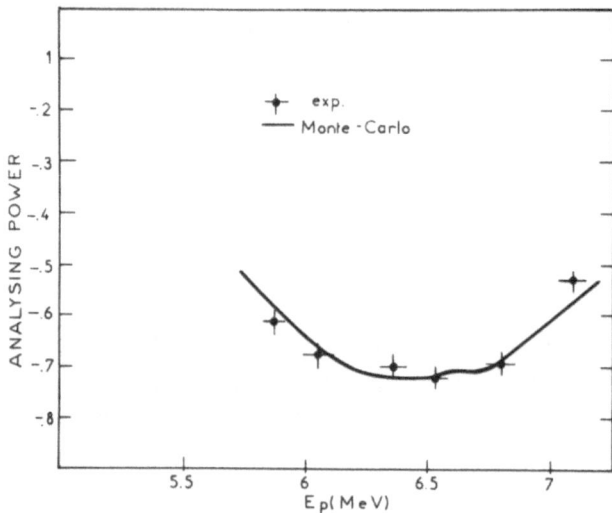

Fig. 7. Analysing power of a polarimeter. The curve represents Monte-Carlo calculations.

5.2. *The Geometrical Correlation*

The geometrical correlation due to the anisotropy of the $^{12}C(p, p)$ cross-section does not allow a direct measurement. Nevertheless, an indirect one can be performed. Formula (2) shows that if the analysing power of one or both polarimeters is zero (P_A or $P_B = 0$) the measured function is reduced to: $F_{mes}(\phi) \simeq C_g \cos \phi$.

We thus replaced one of the carbon scatterers or both by a tantalum scatterer. The proton scattering on the tantalum at low energy being a Rutherford scattering, the analysing power is zero. Since the cross-section of the proton scattering on the tantalum is about 10 times bigger than that of protons on the carbon, it is easy to have a good estimate of the geometrical correlation. Figure 8 shows

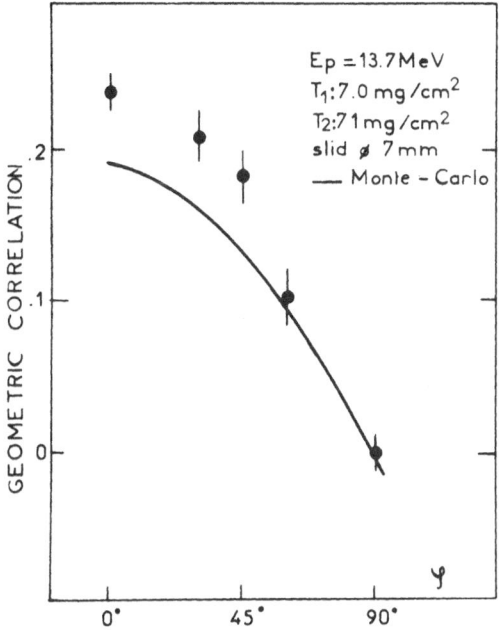

Fig. 8. The geometrical correlation with tantalum scatterers. The curve represents Monte-Carlo calculations. The misfit is probably due to incorrect treatment of the angular straggling in the thick targets.

the values measured for this correlation. We have at $0°$

$$C_{g_{Ta\text{-}Ta}} = 0.23 \pm 0.02 \quad \text{and} \quad C_{g_{Ta\text{-}C}} = 0.127 \pm 0.02;$$

by simple consideration we can deduce the geometrical correlation carbon-carbon

$$C_{g_{C\text{-}C}} = 0.07 \pm 0.025.$$

6. MEASUREMENT OF THE CORRELATION FUNCTION

We measured the correlation function by means of two experiments, each lasting one week. During each experiment the angle ϕ was changed every two hours. Figure 9 gives an example of the distribution of the correlations measured with $\phi = 0°$. We verified that the correlations measured at all angles are dispersed normally around their mean value. Table II gives the correlations obtained.

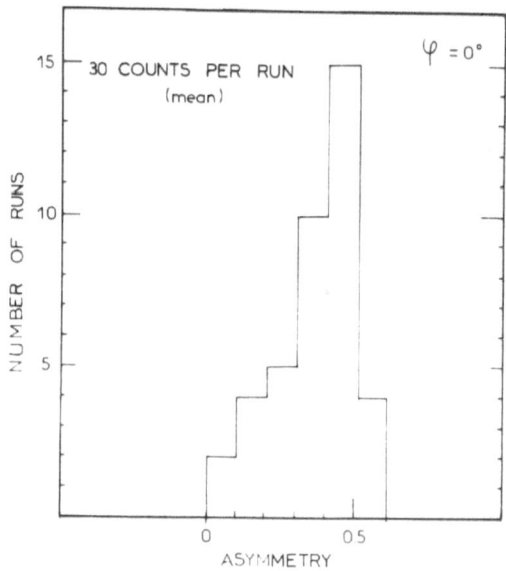

Fig. 9. Histogram of the number of runs in terms of the asymmetry measured for $\phi = 0°$.

TABLE II

ϕ	$D_1G_1 + G_1D_2$ $-D_1D_2 + G_1G_2$	$D_1G_1 + G_1D_2$ $+D_1D_2 + G_1G_2$	$F_{exp}(\phi)$
0°	−673	1769	−0.39 ± 0.025
30°	−440	1626	−0.27 ± 0.025
45°	−478	1863	−0.256 ± 0.023
60°	−258	1489	−0.173 ± 0.025
90°	+15	524	+0.028 ± 0.04

According to the experimental curve of the analysing power, we can give an estimate of $N = \overline{P_A P_B}$. We obtain

$$N_{est} = 0.455 \pm 0.035.$$

With the value of the correlation function measured at 0°, the value of N and the value of C_g, we obtain as experimental value for the correlation function at 0°:

$$C_{nn} = F_{exp}(0°) = -0.96 \pm 0.10,$$

in good agreement with the C_{nn} coefficient interpolated from the measurements of the polarized proton scattering on a polarized target[10], $C_{nn} = -0.945 \pm 0.02$, and $C_{nn} = -0.96 \pm 0.06$ obtained by a set-up similar to ours[9].

The value of C_{nn} of Ref. [10] is more precise than that used to determine N from our measurement at 0°. We obtained

$$N_{exp} = 0.462 \pm 0.033.$$

It is this value which has been used to determine the experimental correlation function with Formula (2). The values of $F_{exp}(\phi)$ thus obtained are compared, in Figure 10, with the estimates of QM and

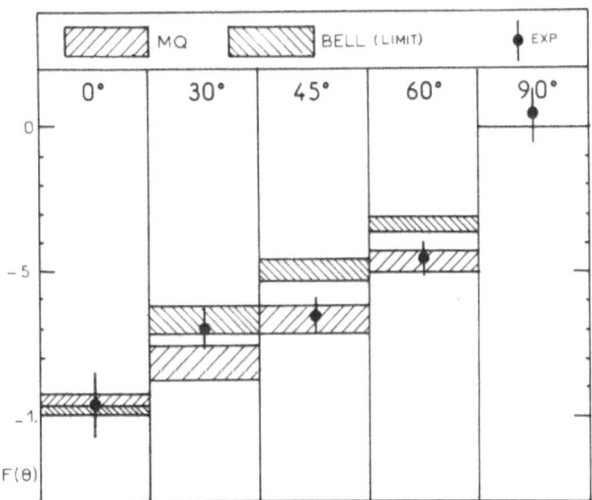

Fig. 10. Comparison of the results of the experiment with the predictions of quantum mechanics and Bell's upper limit.

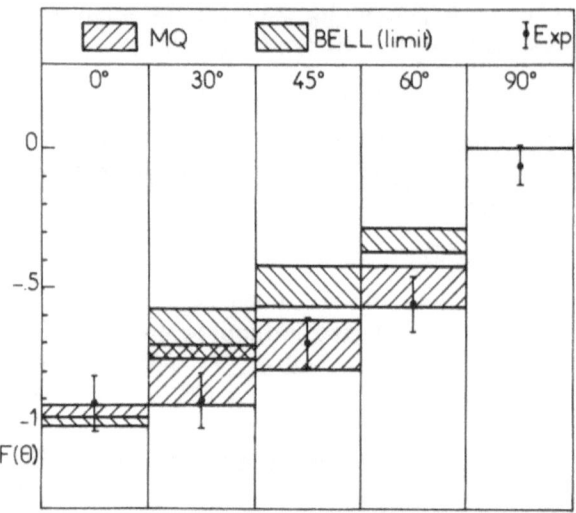

Fig. 11. This experiment has been redone since the Conference with a carbon target of different thickness (18.6 mg cm^{-2}). The results obtained are in agreement with QM and not with Bell's inequality: they are given in this figure.

the upper limit of Bell's inequality. The height of the shaded rectangles represents the uncertainty due to the errors on N_{exp} and C_g.

7. CONCLUSION

At this stage of the analysis, the results of our experiment reject Bell's inequality and are in good agreement with quantum mechanics.

We think a more precise Monte-Carlo calculation (essentially due to a better treatment of angular straggling in the thick target used in this experiment) will allow us to reproduce all the experimental results [12]. We are considering redoing this experiment under different conditions to verify that the conclusions are the same.

Département de Physique Nucléaire,
CEN Saclay, BP 2, 91190 Gif-sur-Yvette,
France

BIBLIOGRAPHY

[1] Wu, C. S. and Shaknov, I., *Phys. Rev.* **77** (1950), 136.
[2] Kasday, L. R., in 'Foundations of Quantum Mechanics', Proceedings of the International School of Physics, Varenna 1970, 'Enrico Fermi' Course 49, ed. by B. d'Espagnat (Academic Press, New York, 1973).
[3] Freedman, S. J. and Clauser, J. F., *Phys. Rev. Lett.* **28** (1972), 938.
[4] Faracci *et al.*, in *Third Int. Conf. on Positron Annihilation*, Helsinki, Finland, August 7–9, 1973.
[5] Holt, R. A. and Pipkin, F. M.; unpublished preprint (1974).
[6] Fox, R., *Lett. al Nuovo Cimento* **2** (1971), 565.
[7] Shimony, A., in 'Foundations of Quantum Mechanics', Ref. 2.
[8] Moss, S. J. and Haeberli, W., *Nucl. Phys.* **72** (1965), 417.
[9] Burman, R. L., *et al.*, C.I.P.N. Paris 1964, Vol. 2, communication 1/C, pp. 180–193.
[10] Catillon, P., Chapellier, M., Garreta, D., *Nucl. Phys.* **B2** (1967), 93.
[11] Terrel, G. E. Jahns, M. F., Kostoff, M. R., Bernstein, E. M. *Phys. Rev.* **173** (1968), 931.
[12] The definitive results have now been published in:
Lamehi-Rachti, M. and Mittig, W., *Phys. Rev. D* **14** (1976), 2543.
Lamehi-Rachti, M., Thesis, Université Paris Sud, 1976.

INDEX

EPISTEME

A SERIES IN THE FOUNDATIONAL, METHODOLOGICAL, PHILOSOPHICAL, PSYCHOLOGICAL, SOCIOLOGICAL, AND POLITICAL ASPECTS OF THE SCIENCES, PURE AND APPLIED

Editor: MARIO BUNGE

Foundations and Philosophy of Science Unit, McGill University

---